Electric Motors
&
Control Techniques
2nd Edition

Irving M. Gottlieb

TAB Books
Imprint of McGraw-Hill

New York San Francisco Washington, D.C. Auckland Bogotá
Caracas Lisbon London Madrid Mexico City Milan
Montreal New Delhi San Juan Singapore
Sydney Tokyo Toronto

© 1994 by **TAB Books**.
TAB Books is a division of McGraw-Hill, Inc.

pbk   5 6 7 8 9 10 11 12 13 DOC/DOC 9 9 8 7
hc   4 5 6 7 8 9 10 11 12 13 DOC/DOC 9 9 8 7 6

Library of Congress Cataloging-in-Publication Data

Gottlieb, Irving M.
   Electric motors and control techniques / by Irving M. Gottlieb. —
2nd ed.
    p.   cm.
   Includes index.
   ISBN 0-07-024011-6 ISBN 0-07-024012-4 (pbk.)
   1. Electric motors—Automatic control. 2. Electronic control.
I. Title.
TK2511.G66 1994
621.46—dc20
                                   93-31288
                                       CIP

Acquisitions editor: Roland S. Phelps
Managing editor: Andrew Yoder
Book editor: Melanie Holscher
Director of production: Katherine G. Brown
Design team: Jaclyn J. Boone, Designer
Cover design: Graphics Plus, Hanover, Pa.

                                                   EL1
                                                   4507

Contents

2 The classic dc motors *32*

3 The classic ac motors *65*

Introduction

FROM SHIPS TO TOYS, FROM STEEL MILLS TO PHONOGRAPHS, AND WHEREVER ELECTRICAL energy has teamed with mechanical motion, the impact of solid-state electronic control of electric motors has made itself felt. It is true that antennas were rotated, tools were driven, and vehicles were electrically powered prior to the advent of thyristors, power transistors, and sophisticated integrated-circuit modules. However, the improvements in precision, flexibility, reliability, and controllability have been so great with the new devices and techniques, that we find ourselves involved with a new and fascinating aspect of technology.

Electric motors, generators, and alternators (the so-called "dynamos" of yesteryear) assumed their roles as industrial "workhorses" during the latter portion of the previous century and the early part of this century. Surprisingly, a perusal of texts dating that far back can still yield useful information about starting, stopping, reversing, and stabilizing such machines. But continued reliance on these venerable methods can only lead to technical obsolescence of machines and techniques. A new era of motor control exerts new demands and, at the same time, stimulates new challenges and provides new opportunities.

When both power engineering and electronics were still in their early stages, those with bold imaginations perceived the potential benefits that might result from a merger of the two arts. A formidable deterrent to such mutuality between these two electrical disciplines was the unreliability of then-available electronic devices and components. During the 1930 to 1950 interim, the electronic control of motors did make some headway as better tubes and components became available for such applications. In particular, *thyratrons* and *ignitrons* attained popularity. It became feasible to electronically control the speed of fractional-horsepower machines and, to some extent, larger integral-horsepower machines. Significantly, some of these circuit techniques are clearly recognizable as the predecessors of present-day solid-state controllers.

This obviously brings us to the solid-state chapter of electronic evolution. Initially, the invention of the transistor sparked a number of application efforts. With the soon-to-follow development of power transistors, the direct control of larger electromagnetic devices became possible. Also, the introduction and quick commercialization of *thyristors* enabled the precise and efficient control of very large motors. Now, a plethora of solid-state devices have become available for arriving a wide range of motor types and sizes; these include silicon NPN and PNP power transistors, N and P channel power MOSFETs, power Darlingtons, germanium power transistors, IGBTs, GTOs, MCTs, and power op amps.

So much for the muscles for motor-control systems. Fortunately, technological-evolution has also provided us with the brains to actuate these muscles in coordinated precision. No end of dedicated ICs have been developed to time, automate, manipulate, and protect these motor-control systems. Both analog and digital techniques are used; excellent control and regulation of motor operational-parameters can be realized without extensive development-projects, or wasteful cut-and-try activities.

The motors, too, have participated in this evolution. The traditional motor-formats, although remaining useful, have to a considerable extent given way to types relying less upon mechanical commutation or conventional single or polyphase power, but much more upon timed-pulses from the alluded "brainy" control-ICs. Such motors exhibit attributes of the long-enduring dc and ac motors, but also possess unique advantages of cost, reliability, and controllability.

The foregoing matters lead very naturally to the final chapter of the treatise dealing with electrically-powered automobiles. Therein, I endeavor to resolve some of the controversies that have long plagued this area of electric-motor application. Interestingly, all of the preceding chapters bear relevancy. The propulsion of the electric auto can be provided by dc or ac motors, and by traditional or new-age motor formats. This stems from the versatility of the aforementioned dedicated control-ICs.

Regardless of specific areas of interest, if you are technically-inclined, you should find useful guidance to practical implementations of electric motors and their control, but I hope that this book will also stimulate rewarding, experimentation and creativity.

The following individuals and firms deserve thanks for their assistance and for their contributions of electronic-control circuits and systems for electric motors: Michael Apcar, President, Randtronics, Inc.; Michael P. Brown, President, ELECTRO AUTOMOTIVE; W.C. Caldwell, Distributor Sales Administrator, Delco Electronics; Walter B. Dennen, Manager, News and Information, RCA; Robert C. Dobkins, Manager of Advanced Circuit Development, National Semiconductor Corp.; Norbert J. Ertel, Marketing Analyst, Bodine Electric Co.; Forest B. Golden, Consulting Application Engineer, General Electric Co.; Alan B. Grebene, Vice President, Exar Integrated Systems, Inc.; Frank A. Leachman, Media Manager, Superior Electric Co.; Larry Steckler, Editor-in-Chief & Publisher, Gernsback Publications, Inc.; Lothar Stern, Manager, Technical Information Center, Motorola Semiconductor Products, Inc.

Irving M. Gottlieb
Redwood City, California

1
Reconsiderations of basic motor and generator action

THE APPLICATION OF ELECTRONIC CONTROLS TO ELECTRIC MOTORS AND GENERATORS has the appearance of a mere merger of two somewhat divergent practices of a common engineering discipline. It is, however, much more than this! It taxes the ingenuity of the practical man and challenges the imagination of the theorist. Indeed, such a merger has evolved as an excellent illustration of applied science. For instance, one might consider such recent innovations as motors with superconducting field magnets, homopolar machines with liquid-metal contacts, magnetohydrodynamic generators, commutatorless dc motors, printed-circuit motors, levitated induction drives for transport vehicles and, of course, the application of solid-state devices to all types of electric machines.

The nature of the new control techniques

As the author, I assume in this book that you have at least a basic knowledge of electrical and electronic devices. Accordingly, I will not attempt to duplicate the contents of the other books already available on electricity, magnetism, and electronics.

This chapter explores the known, with deliberate intent to invoke the unknown. It touches on elemental notions to show how accepted principles band together to produce useful hardware. Basic questions are raised, but the very contemplation thereby initiated will, in *itself*, constitute my objective. From this study, I hope you will realize that electric machine technology, though rooted in the past, is destined for a profuse blossoming in the future.

Let us commence with a discussion of a feature *common to all motors that convert electrical to mechanical energy*—the phenomenon of action at a distance.

Action at a distance

One example of this phenomenon, which greatly perplexed yesterday's scientists, is that of a ferromagnetic object being physically acted upon by magnetic force. Forces other than the magnetic kind also act upon objects or entities separated by a distance. Here, I can think of electrostatic, gravitational, molecular, and nuclear forces. Electromagnetic waves, such as radiant light, might well be included, and there are undoubtably others. For example, those versed in parapsychology often allege that physical objects can, under some circumstances, be acted upon by force fields emanating from the mind. Although the manifestations mentioned are apparently diverse, they incorporate a common feature—*action at a distance without need for an intervening medium*. It was this latter aspect that inspired the postulate of the "ether." Supposedly, the ether was everywhere but it was elusive because of its tenuous nature—it had a viscosity of zero! Although endowed with the properties of "nothingness," it served as the medium of transmission for light and other electromagnetic radiation; that is, it supported wave motion. The static forces exerted by magnets, by charged bodies, and by gravitation were not so glibly explained. Then, and now, such forces were simply ascribed to "fields" and the role of the ether was somewhat more nebulous.

The mathematicians next extended the concept of the field and endowed it with properties of *self-propagation*, which eliminated the need for the ether. This new concept came at a favorable time, because experiments carried out to detect the presence of the ether were not successful. Besides, whether one comprehended the mathematics or not, the notion of radiant energy traversing the vacuum of space did not stretch the imagination any farther than the elder hypothesis. Fortunately, the mathematical descriptions provided by Maxwell's equations and by other theories dealing with fields, harmonized with experimental investigations and facilitated the development of practical devices. Nonetheless, the imaginative mind remains undecided over the part played by the intervening space when any type of force exerts influence over a distance. If the influence manifests itself over a gap of true "nothingness," does this imply the propagation of particles of some sort between the source of the action and that which is acted upon? This, too, was given much thought by many brilliant minds. And, like the ether, we find that the emitted particles are still with us, but dressed in a new style. For example, the prevailing concept of the *photon*, the elemental carrier of radiant energy, postulates a rest mass of zero. Thus, the ghost of the extinct ether returns to haunt us!

There is considerable scientific speculation that gravity, magnetism, and the electric field are somewhat different manifestations of a *universal* law of nature. The coupling between magnetism and moving electric charges is evidenced by electric motors and generators and by the myriad devices that exploit the phenomenon of electromagnetic induction. It is easy to assume a matter-of-fact attitude regarding the relationships between electricity and magnetism. However, it is instructive to reflect that these "simple" facts of technological life eluded the scientists of the nineteenth century until they were experimentally observed and interpreted. In one case, Hans Christian Oerstead recognized the significance of the deflection of a compass needle by a current-carrying wire. In the other case, Michael Faraday was seeking a relationship between *steady* magnetic fields and electric currents in stationary conductors. Although he found none, he recognized the significance of currents induced in certain situations where relative motion existed between the magnetic field and the conductor.

In our era, the assumed relationship of gravity to electricity and magnetism has thus far been quite elusive. Despite the powerful concepts of *relativity* and *quantum physics*, the relationship between gravity and other forces does not appear to be strong. But perhaps history will repeat itself—a surprise observation might one day be made of an unsuspected cause-and-effect relationship that will shed a new light on the nature of gravitational force. Actual experiments are already being carried out to detect and explain "gravity waves." It is known that gravitational force does not communicate its influence instantaneously. Like the forces of magnetic and electric fields, gravitational influence cannot propagate through space faster than the speed of light. And the force associated with the gravitational field, like that of the magnetic and electric fields, diminishes inversely in proportion to the square of the distance between two bodies, poles, or charges. Unlike the force fields of magnetism and electricity, gravitational force produces only attraction, never repulsion between bodies! Although this subject has been a favorite for science-fiction writers, the quest for a method of reversing, or neutralizing, gravity is by no means the exclusive indulgence of those who deal in fantasy.

Just as practical motors and generators have been profoundly influenced by research in cryogenics and superconductivity, solid-state theory, plasma physics, and materials technology, subsequent progress in the development of a unified field theory can be expected to manifest itself in improved hardware and new control techniques. Surely, the harnessing of basic forces is what motors and generators are all about!

Electrostatic force

Pretend that our electrical technology exists and that you have a good grasp of it but that, somehow, electric motors have not yet been developed. Given the assignment to create such a device, how might you proceed? A reasonable way would probably be to investigate forces capable of acting on physical materials of some kind. Then, you would think of some way to produce *torque*, or a *turning motion*. This would be an encouraging step, but the torque would also have to be continuous, so that a constant rotation would ensue. However, the mere attainment of this objective might not result in a practical motor. For example, the well-known novelty item, the *radiometer* (Fig. 1-1), converts the energy of incident light photons to kinetic energy

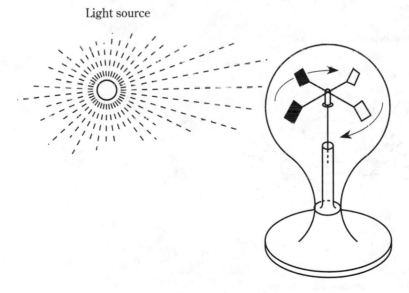

1-1 The radiometer.

by the rotation of its windmill-like blades. But only a feeble torque is developed in such a device. It would be futile to provide the radiometer with a shaft so that work could be performed. Perhaps such an approach might be put aside reluctantly because certain instrumental applications could be visualized. But for use in the environment, this type of "motor action" does not appear promising.

You might next consider electrostatic force. Figure 1-2 displays the field patterns of point charges. Figure 1-3 shows an *electrostatic voltmeter*. Here, the at-

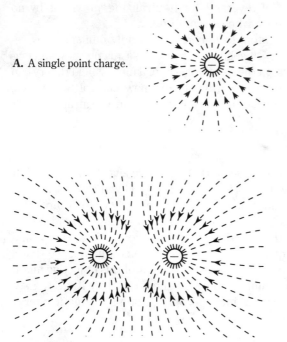

A. A single point charge.

B. Repulsion between similar charges.

1-2 The field pattern associated with electrostatic lines of force.

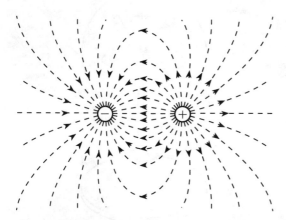

C. Attraction between dissimilar charges.

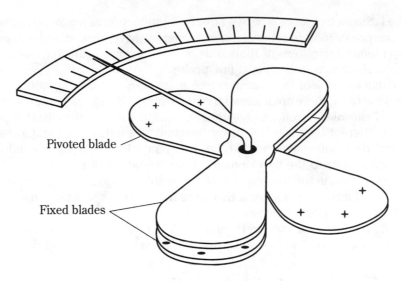

1-3 The electrostatic voltmeter.

traction of unlike charges is evidenced as usable "motor action" in this device. To be sure, the rotation of the electrostatic voltmeter is not continuous, but perhaps this can be arranged. Of course, the torque developed by this meter movement is still woefully inadequate for the needs of industrial motive power. Maybe you would ponder whether an electrostatic motor could be devised to develop the turning power needed.

Consider a current of *one ampere*. Such a current is readily produced by small batteries and is safely carried by ordinary 18-gauge hookup wire. Now, one ampere represents the flow past a point of approximately 3×10^9 electrostatic units of charge per second. Because the charge of the electron is $4.8 \times 10_{-10}$ electrostatic units, it follows that one ampere corresponds also to 6.25×10^{18} electrons per second. This number of electrons is defined as the *coulomb*, so finally we say that one ampere of current flows in a circuit when the rate of charge is one coulomb per second. Apparently, the coulomb is *not* a wild concept described by fantastic numbers. In many ordinary electrical and electronic devices, you can expect to deal with currents ranging from several tenths to several tens of coulombs per second.

Using Coulomb's law, it is easy to show that if two metallic spheres, one centimeter in diameter and separated by one meter, center to center, could somehow be oppositely charged with one coulomb of electricity, they would develop the fantastic attractive force of approximately *one million tons*. The conditional "somehow" is well used, for the potential difference developed by such an electrified system would be in the hundreds of teravolts. Long before such an astronomically high voltage could be brought into existence between the spheres, a cataclysmic lightning flash would have disintegrated the apparatus. If you scale down the charges and alter the geometrical configuration of the spheres, or plates, the best that can be accomplished falls miserably short of what is easily, compactly, and economically achieved when the basic motive force is derived from the interaction of magnetic fields. Fractional-horsepower motors using electrostatic forces have been experimented with, but they require many tens of thousands of volts, involve critical insulation techniques, and show little indication of practicality.[1]

Although the idea of electrostatic force as the torque-producing source for motors was dispensed, there are many who remain intrigued with the fantastically powerful force fields of electrons. Among these people, the feeling prevails that perhaps a radically different technique might yet be found to use this elemental force of nature to directly produce mechanical rotation at power levels suitable for industry.

A possible spur to the development of electrostatic motors is the great stride that has been made in the transmission of very high dc voltage. Also, there are whole new families of insulating materials that were nonexistent some years ago when interest declined in electrostatic motive power. Of considerable relevance, also, is the recently attained state of vacuum technology. The future might hold some interesting surprises for those who view electrostatic motive power as a dead issue.

Consideration of the use of magnets to achieve motor action

The general practical aspects of magnets are now well known; the considerable forces of attraction and repulsion evident between the poles of strong magnets naturally suggest the possibilities of motor action. Not only do ferromagnetic materials have force fields available with reasonable shapes, they have *permanent* magnetization as an added dividend. However, if you did not know otherwise, it would be easy to cite reasons why the magnetization of magnets should be "used up" under certain circumstances.

Magnetic lines of force surrounding bar magnets are shown in Fig. 1-4. The magnets behave as if such field lines had the following characteristics:

- Like poles repel; unlike poles attract.
- The forces of attraction and repulsion are the same when pole strengths, distances, and arrangements are the same.
- Pole pairs always exist in a magnet in such a way that one might say that the lines leave from a north pole and enter at a south pole. *There are no unipoles in practical magnets*.
- Lines of force *never cross*. In a space subjected to fields from more than one source, a *resultant field* is produced, having a density and direction determined by the directions and strengths of the contributing fields.
- In the case of repulsion, it is more correct to speak of field deflection than of neutralization. In other words, lines of force have their paths altered, but they are not destroyed.

---

[1]Benjamin Franklin actually made an electrostatic motor! It produced rotation from the stored electrostatic energy in a Leyden jar. A modernized version of this machine developed one-tenth horsepower when powered from a 30,000-volt electrostatic generator. Electrostatic motors must also consume current to produce useful torque. The situation remains, as in Franklin's day, tantalizingly suggestive, but generally impractical.

- The lines of force surrounding a magnet mutually repel one another.
- Lines of force have been likened to rubber bands in that they seek the shortest path. A somewhat better statement would be that they seek the easiest magnetic path.
- Forces arising from magnetism obey the inverse square law.

Interestingly, there have been reports of discoveries that the magnetic field has a *particle* nature. The search for the elusive *magnetic monopole* has apparently been rewarded with success. Although this can have tremendous ramifications for motor technology, the nature of practical implementations is not yet discernible.

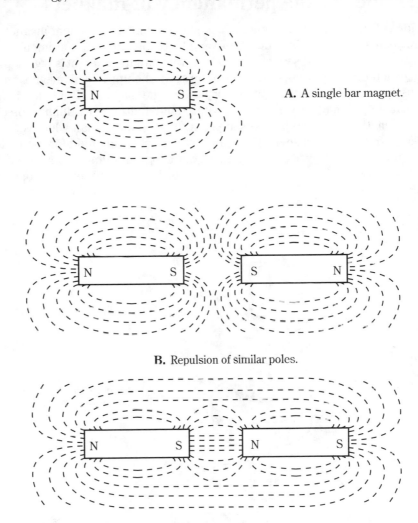

A. A single bar magnet.

B. Repulsion of similar poles.

C. Attraction of dissimilar poles.

1-4 The field patterns associated with magnetic lines of force.

All magnetic action is believed to stem from the effect of *charges in motion*. Most often involved is the orbital and spin motion of electrons in atoms. In most atoms, the electrons are paired off, so no external magnetic effect occurs. But in the atoms of ferromagnetic materials such as iron, nickel, and cobalt, there is a net magnetic moment due to the spin of electrons not paired with other electrons having counterspins. In a sense, there are no "natural magnets"—magnetism is the result of charges in motion. (Remember the "forces" produced by a gyroscope in which gravitation like fields result from the rotation of a mass.)

Examples of the permanency of magnets

The idea of "using up" the magnetic field surrounding a permanent magnet has already been alluded to. Consider, for example, the levitation experiment depicted in Fig. 1-5. A common-sense approach to this situation might suggest that the price exacted for resisting gravitational force on the upper disk magnet would be the dissipation of magnetism in one or both of the magnets. But, such is not the case. The upper magnet will remain levitated indefinitely. Any diminution of magnetic strength in the two magnets appears to be no more than what would occur if they were stored on a shelf. This simple situation is truly food for thought—students in beginning physics often present quite logical arguments suggesting that the magnets *must* become demagnetized in order to satisfy energy conservation laws!

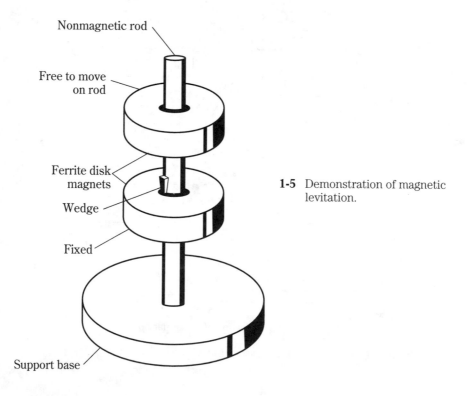

1-5 Demonstration of magnetic levitation.

A somewhat more sophisticated experiment is illustrated in Fig. 1-6. Here, the magnet provides the field for a basic alternating-current generator. Not shown is some kind of a prime mover that rotates the armature loop and slip-ring assembly, which are on a common shaft. (A small gasoline engine would serve this purpose.) The electrical load is a light bulb; its glow provides visual evidence of the production of electrical energy. Again, the question might be asked whether the magnet is thereby "used up"; that is, whether it is demagnetized in exchange for energy extracted from it. Now that you are familiar with these matters, you might correctly suspect that such is not the case. But, if you were contemplating the original design of such a device, you would likely feel concern over such a possibility. In actuality, the energy exchange occurs between the *mechanical* energy supplied by the prime mover and the electrical energy delivered to the load.

1-6 Elementary ac generator.

Attempts to produce motor action from permanent magnets

By experimenting with horseshoe and bar magnets, you can see that forces of both translation and torque can be produced. It could naturally be inferred that *continuous* rotation might ensue from an appropriately arranged assembly of magnets. A possible setup is shown in Fig. 1-7. Assume that thrust bearings are associated with the ends of the shaft so that the two sectors can neither come together nor fly apart.

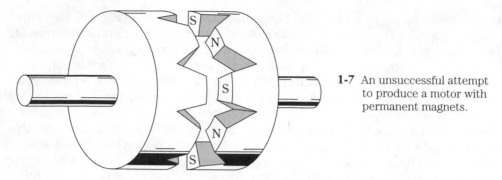

1-7 An unsuccessful attempt to produce a motor with permanent magnets.

No matter how the two sectors are magnetized, the best result that can be achieved is a *transient* rotary motion, after which the sectors lock in position. No one has yet succeeded in producing continuous rotational motion from the interplay of force fields from permanent magnets alone! However, such failure is not due to the same reason why perpetual motion cannot be brought about. It is, rather, a matter of "switching logic." At an appropriate instant in the momentary turning motion, something would have to be done to prevent the oncoming condition of "lock-up." Devising such an arrangement could be an interesting project for long winter evenings.

The electric charge situated in a magnetic field

Thus far, in our imaginary search for motor action, we have come tantalizingly close to the desired result. The interaction of electric charge fields provided such food for thought. Tremendous forces are associated with charge accumulations, but so are tremendously high voltages. Charges measured in tiny fractions of a coulomb still involve tens of kilovolts of potential difference, but they yield motor forces (torques) of not many tens of ounce inches. The electrostatic voltmeter, although incapable of continuous rotation, was cited as an example.

The interaction of magnetic force fields is also stimulating to the imagination, because forceful movements could be produced without any danger of disrupting the air. (Extremely powerful magnetic fields produced by superconducting magnets still impose no breakdown mechanism on nonmagnetic or paramagnetic substances, such as the gasses comprising the atmosphere.) It was pointed out that there is apparently no way in which continuous rotation can be brought about by any arrangement of permanent-magnet poles. This is frustrating, but the immense forces available from even small magnets make it difficult to part with the notion of using magnetic fields. In view of these situations, our investigations might next be expected to focus on interactions between electric charges and magnetic fields. Perhaps, the best features of both force fields can thus be made available.

A good starting point would be the evaluation of force imparted to a fixed unipole of electric charge situated in the interpolar field of a strong permanent magnet. The charge could be on a small aluminum sphere suspended by a nylon thread. If the sphere were slowly lowered into the magnetic field, exactly nothing would happen. The purist might object to this statement because the electric lines of force would no longer remain nice and symmetrical as depicted in Fig. 1-8. However, the distortion

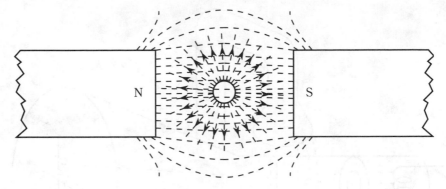

1-8 Stationary magnetic field does not affect the electrostatic field of a fixed charge.

of the electric field would be no different than that produced by a "dummy magnet," say a soft-iron structure that had been completely demagnetized. And while the charge was being lowered into position, the charge would be moving, rather than static. But, if this were done slowly, it would not alter the basic idea revealed by the experiment— that a fixed charge situated in a *stationary* magnetic field *experiences no mechanical force*. In the performance of this experiment, it is desirable that the sphere be small, that the magnet poles be widely separated, and that the sphere be carefully lowered midway between the poles. (Otherwise, the sphere might move to one pole or the other because of electrostatic attraction.) Thus far, our hoped-for merger between electric and magnetic forces shows no prospect for realization.

Successful demonstrations of interaction between electric charges and magnetic fields

Having failed in the initial attempt to merge the effects of electrostatic and magnetic forces, it's time to go back to the books. After some perusal and contemplation, you might produce two instances of motor action derived from the interaction of charges and magnetic fields. In both instances, charges are deflected when immersed in a magnetic field. In the first situation, illustrated in Fig. 1-9, the electron beam of a cathode-ray tube is deflected in the horizontal plane when vertically oriented magnetic lines of force are introduced. To be sure, this is not yet our desired goal of producing a force in a solid object, but it is nevertheless an encouraging experiment because it demonstrates that an interaction between the electric and the magnetic field can indeed be achieved.

The second situation, shown in Fig. 1-10, demonstrates the Hall effect. Here, a difference of potential occurs across the edges of a strip of conducting material in response to the simultaneous application of voltage and a magnetic field, which are space oriented as shown. The basic idea is that, *without* the magnetic field, *no* potential difference would be monitored by the microammeter. This is because the connections of the microammeter are at equipotential points with respect to the flow

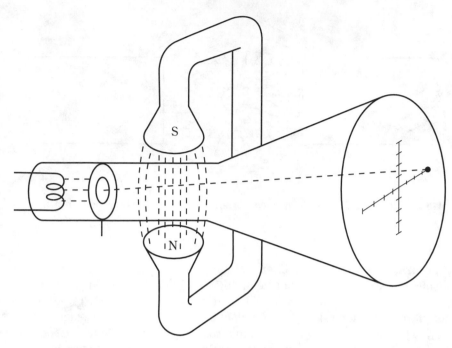

1-9 An example of motor action imparted to charges in a magnetic field.

of the main current, that is, the current produced by the battery. In essence, the microammeter is a null detector connected to opposite points of a balanced Wheatstone bridge. Assuming that the conducting strip is of homogeneous material, and that the microammeter is connected precisely at the center line dividing the length of the conducting material into two equal parts, you would not expect to monitor a difference of potential at right angles to the flow of the main current.

Does the introduction of the magnetic field alter the uniform conductivity of the strip? This, of course, could be offered as an explanation, and, in a sense, it is correct. However, it is the mechanism by which this apparent effect occurs that is interesting. Specifically, it comes about because of the "motor" effect of the magnetic lines of force on the electron drift, which is the main current through the strip. These drift electrons are deflected as shown in Fig. 1-10. Because of their resultant density at one edge of the conducting material, and their relative scarcity at the other edge, a difference of potential appears across the faces of the material to which the microammeter is connected. (What appears at first to be a demonstration of generator action turns out to be also an example of motor action.) Interestingly, the mutually perpendicular relationships of charge flow (electric current), magnetic field, and deflection are common to *both* experiments. Moreover, the *directions* of these quantities are also consistent in both phenomena. In summary, progress was made because it is clear that *moving* charges experience a deflection in a magnetic field.

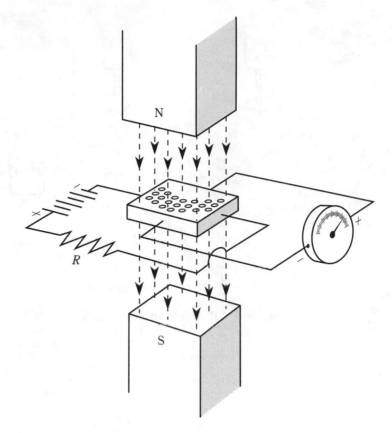

1-10 A demonstration of the Hall effect.

The physical deflection of current-carrying conductors

The Hall-effect experiment is suitable for demonstrating direct motor action on a physical object—the conducting strip itself. However, the assumption was made, even though not emphasized, that the main current through the strip was too low to produce a readily observable force. This, aside from making life easy on the battery, was the function of the current-limiting resistance, R. Here is a good example of coming very close to an important phenomenon, yet failing to discover it.

Inasmuch as experiments have indicated that interactions do exist between moving charges and magnetic fields, one could logically be led to such an experiment as that shown in Fig. 1-11. The rationale for the setup would be that the electric current through wires involves *both* moving charges and magnetic fields. The magnetic field surrounds the wire in the form of circular lines of force, as depicted in Fig. 1-12. The lines of force are concentric and become more and more dense as the center of the wire is approached. This notation tells us that the magnetic field strength increases as one progresses from an outside point toward the conductor.

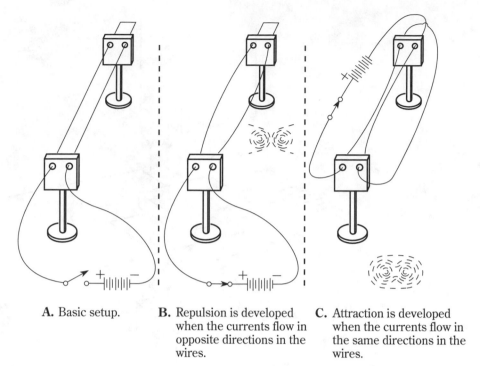

A. Basic setup.

B. Repulsion is developed when the currents flow in opposite directions in the wires.

C. Attraction is developed when the currents flow in the same directions in the wires.

1-11   Motor-action experiments in which actual physical displacements are produced.

In Fig. 1-11A, the basic setup is illustrated. The parallel wires are assumed to be stranded for flexibility and to have sufficient slack in them to allow for movement. Indeed, upon closing the switch in Fig. 1-11B, the wires actually move apart from one another when subjected to the *repulsive* action of adjacent magnetic lines of force. And, when the current is made to flow in the *same* direction through both wires, as is the case in Fig. 1-11C, the wires, impelled by *attractive* force, move together. The actual physical displacement of copper wires is a big leap forward from the mere displacement of charges in a vacuum, or within the confines of conducting material. Significantly, the physical movement of the wires can be interpreted by two equivalent ways of looking at the phenomena. On one hand you could deal with the interaction between the magnetic field of one wire and the moving charges of the other wire. On the other hand, you need merely consider the effects of the adjacent magnetic lines of force from each wire.

The mechanical forces responsible for the deflection of the wires do *not* stem from electrostatic effects. The external surfaces of the wires are neutral as far as electric fields are concerned. This can be proven by inserting a sheet of conducting material, such as aluminum, between the wires. When such "shielding" is provided, the current-dependent movements of the wires remain unaffected.

An important aspect of the current-carrying conductor is that the conductor experiences *no* mechanical force from its own magnetic field. This may or may not be

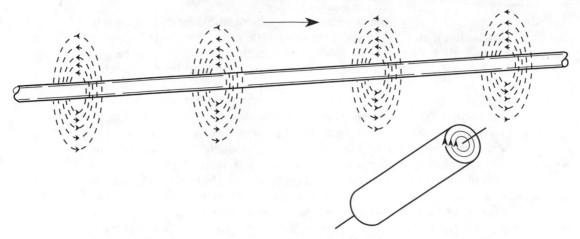

A. Perspective view. The field pattern is actually of a continuous cylindrical form as shown in the insert. However, concentric circles convey the essential information.

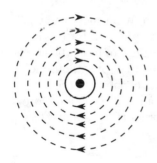

B. "Head-on" view of current-carrying conductor. Inset depicts more realistic view of field density.

1-12   The magnetic field surrounding a current-carrying conductor.

obvious. If the current in a conductor is suddenly broken, a counter EMF is produced in the conductor from the action of its own collapsing magnetic field. Thus, a current-carrying conductor in conjunction with its *own* magnetic field can be likened to a generator. However, motor action is only forthcoming when there is interaction between the current-carrying conductor and an independently produced magnetic field. The latter, however, can be produced by another portion of the conductor, as is the case in the setups of Fig. 1-11.

The solenoid as a means of manipulating the magnetic field

Having found that genuinely macroscopic forces are available from the magnetic-field interactions of current-carrying conductors, the quest for practical motor action would, nevertheless, probably be doomed to failure. Although you could dream up ingenious ways to reverse currents, and therefore the direction of magnetic fields, motors based on the interaction between two single conductors would have features best described as toylike. The geometry of such an arrangement would be embarrassingly awkward, and such a motor would probably serve little more than to pull its own weight. What is needed is a means by which the magnetic lines of force can be *collected and focused within the confines of a small area or volume*. With such a technique, you would have at your disposal a relatively strong magnetic field in a compacted space. This is because the field strength is determined by the density of the magnetic lines of force. Thus, in a bar magnet, the so-called poles are those regions external to the bar itself, where the density of the magnetic field is the greatest.

You might experiment with the idea of paralleling conductors, and indeed, this would be a step forward. Consider the solenoid, an arrangement in which the conductor consists of a layer of coiled wire, or many such layers. In Fig. 1-13A, the collecting and focusing action of the solenoid is shown. In its ultimate effect, the current-carrying solenoid closely duplicates the behavior of the bar magnet as far as the external magnetic field is concerned. This, indeed, led to early notions of magnetism as a phenomenon stemming from the overall effects of a tremendous number of tiny current loops. At first, this was attributed to the orbital motions of certain electrons, but later investigations suggest electron spin as the basis for magnetism in ferromagnetic materials.

The solenoid provides a *controllable* source of magnetism. What more can you desire in your search for techniques suitable for producing motor action? It turns out that you need even *more powerful* concentrators of magnetic lines of force than the air-core solenoid. This is true even though you might be willing to deploy thousands of turns and to utilize current magnitudes up to the maximum capacity of the wire. Essentially, you want to retain the basic action of the solenoid, but to obtain more magnetic lines of force per ampere. Fortunately, the greed for more output and less input can be easily satisfied.

The iron-core solenoid

If low-carbon iron or other ferromagnetic material is inserted in the solenoid, the situation depicted in Fig. 1-13B will result. The density of the magnetic lines of force emanating from, and returning to, the polar regions is now greatly increased. In other words, a *stronger* field is available for the production of either motor or generator action in rotating machines. So-called soft magnetic material is generally used so that initial current flow will not produce a permanent magnet. "Hard" magnetic materials would result in high magnetic retentivity and would deprive the solenoid of its electrical *control* feature. Generally, the strength and polarity of the magnetic poles is desired to be a function of the current flowing in the winding. Additionally, the use of a "hard" ferromagnetic material would result in inordinately high hysteresis loss when such a solenoid was energized by alternating current. Magnetic hysteresis is a friction-like phenomenon deriving from the sluggish efforts of magnetic domains to continually realign themselves in response to the changing field produced by an alternating current. (Each domain comprises millions of atoms with net magnetic fields arrayed in the same direction.)

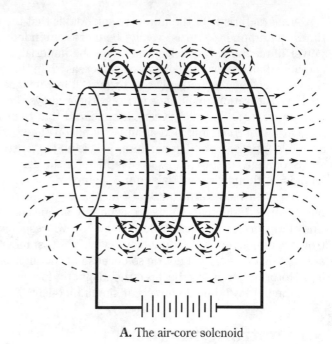

A. The air-core solenoid

1-13 Increasing the magnetic
field strength of a
solenoid.

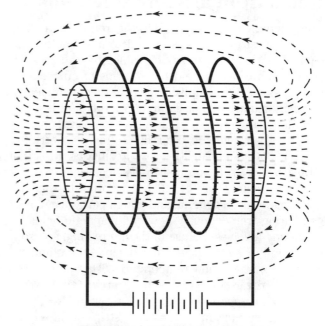

B. Same solenoid with core of iron
or other ferromagnetic material.

A characteristic of the iron-core solenoid is that the magnetizing force (current times the number of turns) cannot be increased indefinitely in the quest for more-powerful magnetic fields. Depending on the material, an area of diminishing return is reached as the magnetizing force is increased. In this area of operation, the material is said to be magnetically *saturated*. All ferromagnetic materials are characterized by saturation, including ferrites, their alloys, and even those that actually contain no iron. For many years, it appeared that the saturation behavior of ferromagnetic materials imposed an upper limit on the strength of the magnetic fields available for motors and generators. This is no longer true.

Seemingly, nature has exacted a price for the gains accompanying the use of solenoids. Although magnetizing force (ampere-turns) increases *directly* with the number of turns, it turns out that *inductance* increases as the *square* of the number of turns. Inductive effects are generally of nuisance value in rotating machines. They interfere with commutation and slow down response time when electronic controls are employed. A dc shunt motor, which at first might be expected to operate on alternating current, will not do so because of the high *inductance* in its field winding. Motor action is lost because the magnetic forces from the armature and those from the field winding do not occur simultaneously.

Superstrong magnetic fields from air-core solenoids

The magnetic field developed by the iron-core windings, or electromagnets, of rotating electrical machines is limited to a strength of about 15,000 gauss (lines per square centimeter). Above this value, the iron goes into saturation, and greater field strengths are dearly paid for in terms of current and temperature rise. Although much has been accomplished in materials technology, a ferromagnetic material that has significantly higher saturation levels and is also economical and workable for manufacturing purposes has not yet appeared on the horizon. Because higher magnetic-field strengths lead directly to more powerful motors and to generators with greater output capabilities, any technique for attaining superior magnetic-field strength is of prime interest to the electrical-machinery industry.

After many years of quiet resignation to the saturation behavior of ferromagnetic materials in general, and iron in particular, a new and sophisticated method of generating tremendously powerful magnetic fields has come into being. This new technique involves the use of the interesting, and somewhat mysterious, phenomenon of *superconductivity*. In principle, the ohmic resistance of many metals vanishes at extremely low temperatures. That being the case, solenoids can be wound with little regard for cross-sectional area of the current-carrying conductor. Because current flow encounters negligible resistance at these extremely low temperatures, gigantic currents can be carried by strips or films of appropriate material. The most significant aspect of the new art is that iron or other ferromagnetic substances are not used—the super-magnetic fields are produced in *air*. And the air neither saturates nor becomes involved in any breakdown mechanism.

The brittle intermetallic nobium-tin compounds have been used a lot in superconducting magnets. By means of vapor deposition or by a diffusion process, the compound is formed on a supporting strip of a more ductile material. By similar techniques, alumina or an organic coating is applied as an insulating film. A solenoid wound with this composite material is then immersed in a liquid-helium bath. Superconductivity commences quite abruptly in the vicinity of –260°C.

Successful motors and generators have been constructed using superconducting fields in the neighborhood of 60,000 gauss. Such machines have included 150-kW generators and motors with output capacities exceeding 3000 horsepower. Moreover, it appears practical to make superconducting electromagnets with field strengths of over 140,000 gauss. Such magnets need be only about 15 cm in diameter and operate continuously from a dc source of a few hundred watts. Cryogenic techniques, material technology, and experimental discoveries are advancing at a hectic pace; it would be difficult to impose either upper or lower limits on the parameters associated with superconductivity. A question not yet answered is: *Can a material be produced that will superconduct at "ordinary" temperatures?*

The controllability of coil-produced fields leads to a true motor

An example of motor action obtained with the use of air-core solenoids, or coils, is the electrodynamometer-type wattmeter illustrated in Fig. 1-14. In the arrangement shown, the moving coil carries a current proportional to the voltage of the power source (or very nearly to the voltage impressed across the load). This coil consists of many turns of fine wire with a current-limiting resistance generally inserted in series with it. The two stationary coils are connected in series-aiding and also in series with the load. These coils comprise relatively few turns of heavy-gauge wire. The moving coil experiences a torque closely proportional to the product of the voltage across the moving coil and the current through the stationary coils, thereby producing a deflection determined by the product $E \times I$, or watts. Because the angle of the interacting fluxes changes with the physical rotation of the moving coil, a slightly nonlinear scale is necessary. (Not shown are spiral countertorque springs, which tend to restore the moving coil-pointer system to zero.)

The test functions of this device are not of direct interest to us, but it is quite clear that *motor action* is achieved. To be sure, the rotation is limited to a small arc, rather than being continuous. But, the extension of this operation to continuous rotation should not require a high degree of inventiveness. It is also apparent that this type of motor action should be obtainable from an ac source as well as from a dc source. This is because the magnetic-field directions of the moving coil and of the stationary coils both change together. Therefore, the torque developed remains unidirectional. Indeed, such a wattmeter can be used in both dc and ac circuits.

The commutator-type dc watt-hour meter of Fig. 1-15 is the logical extension of the wattmeter just discussed. Here, is our first true motor, inasmuch as continuous rotation is obtained. The shaft of this motor drives a mechanical readout. In turn, the rotation of the shaft is restrained by countertorque developed in the eddy-current disk. However, these shaft-attached devices relate most relevantly to the instrumental purposes of this "motor." From the standpoint of continuous rotation, the significant component is the commutator and brush assembly. In simplest terms, this item

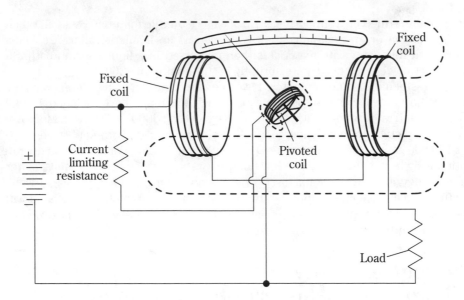

1-14 The electrodynamometer-type wattmeter.

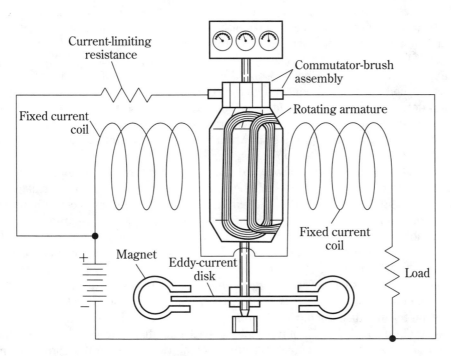

1-15 The commutator-type watt-hour meter.

is a rotary switch that automatically reverses current flow in the armature so that a unidirectional torque, rather than a "hangup," is developed. And because the commutator is mounted directly on the shaft, the current-reversing switching cycles are synchronized with the turning of the armature coils.

Another source of motor action

The experimental setup illustrated in Fig. 1-16, although of primitive nature, involves important ramifications for motors. In this discussion, suppose that material "X" is:

    (a) An insulator.
    (b) A perfect conductor.
    (c) A very good conductor.
    (d) Iron or other ferromagnetic substance.

In the first instance (a), when the material is an insulator, such as a sheet of glass, Lucite, mica, etc., nothing at all happens—at least physically. The purist might postulate transient phenomena with regard to charges, but such disturbances are of no practical importance to motors. Indeed, sensitive and complex instrumentation

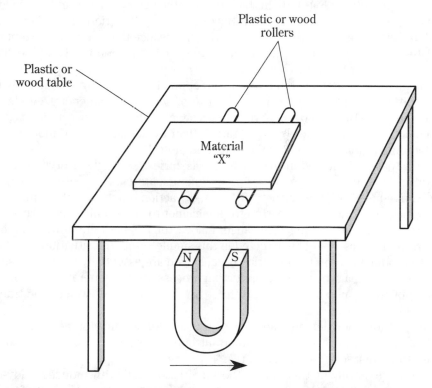

1-16 A basic experiment with important implications for motor technology.

would have to be employed to detect any such cause-and-effect relationship. In this discussion, the interest is in gross physical movement. That is why the sheet of material, X, is placed on rollers. For all practical purposes, the sweeping magnetic field produced by the movement of the magnet produces no electrical or magnetic effects capable of setting material X in motion.

In situation (b), a perfect conductor is used for material X. Suppose that this is superconducting material without regard to the awkward implementation of such an experiment. You have, of course, complete liberty to imagine such an investigation. It happens again that the sheet is subjected to no physical displacement. This is because a perfect conductor does not allow penetration of currents induced by a time-varying magnetic field. This self-shielding property is not entirely foreign—a similar occurrence was encountered under the name "skin effect." In principle, an ideal conductor that is also not ferromagnetic will experience virtually no mechanical force in this experiment.

If the material is a very good conductor (c), say copper or aluminum, it will move in the direction of the moving magnet. Such material will have induced in it so-called eddy currents. By Lenz's law, the magnetic fields of these eddy currents will oppose the inducing field and will thereby *carry* the material away from the advancing magnet. This is the very same principle involved in the eddy-current disk associated with the watt-hour meter in Fig. 1-15. In this situation, the magnet moves and the conductive material is initially at a standstill. But the *relative* direction of the forces produced in the two cases are the same. Knowing in advance that the eddy-current disk generates a retarding torque, imagine the magnets to be in motion and that the disk is initially at a standstill. A little contemplation reveals what must happen in such a case.

The disk would actually rotate in the direction of the "orbiting" magnets. With a little additional imagination, envisage the use of *stationary* ac-energized electromagnets in place of the moving permanent magnets. Then, the sweeping magnetic field could be produced *electrically*, rather than by actual physical motion. And, thereby brought into existence is the induction motor!

The situation indicated in (d) is easy to visualize. Seemingly, everybody is able to deduce the effect produced with ferromagnetic material. As in situation (c), the sheet would move with the magnet. However, the action is not quite the same. In (c) a quick movement must be imparted to the magnet to get a somewhat delayed motion of the sheet material. In (d) the sheet follows the magnet right down to zero velocity. Although this would be detected in this simple setup, the situation in (c) also tends to produce heat in the material, because you are essentially inducing a short-circuited current path by the same induction process used in generators. Finally, the situation of (d) results in a downward force on the material, as well as the translatory motion.

The phenomenon demonstrated in (d) is useful for certain clutches, brakes, and coupling devices. However, the all-important implications with regard to motor action are those derived from situation (c).

Another experiment that shows the force associated with induced eddy-currents is shown in Fig. 1-17. The copper disk is, in essence, the physically free but shorted secondary of a transformer. It goes without saying that actual transformers are also subject to such disruptive force. Indeed, in large transformers, the designer

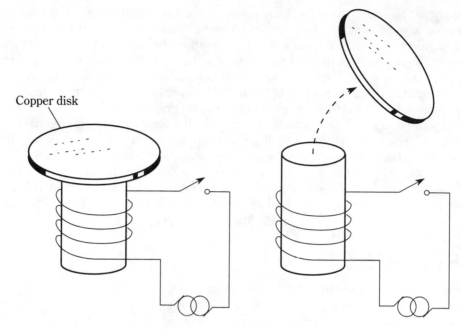

Copper disk

A. Conductive copper disk is placed on nonenergized solenoid.

B. Solenoid is energized from ac source.

1-17 Another demonstration of eddy-current induced forces.

must give as careful attention to the *physical* integrity of the windings as to purely electrical or magnetic considerations. This experiment could be performed with a dc source, in which the repulsive force developed would be of a transitory rather than a sustained nature. The use of ac relates the experiment more closely to the *induction motor*.

With regard to both motor action and generator action, technical literature continues to perpetuate confusion about that old bugaboo—the direction of current. This is particularly noticeable when you compare the left-hand and right-hand rules in different books. These "rules" enable you to determine the directions of magnetic fields, mechanical force, and current by assigning these parameters to different fingers of a specified hand. Usually, a dot signifies that current flows out of the page, and a cross indicates that current flows into the page. There have been two opposite conventions based on the way the electrical current is supposed to flow. It is customary for texts to state that it makes no difference which convention is adopted as long as one is *consistent* in using the selected one. This assertion is true, but confusion nonetheless continues because of a persevering sloppiness in semantics that continues unabated in motor books.

Specifically, if you choose to deal with the older concept that electric current flows from the positive terminal of a battery or other active sources, through the circuit, and thence back to the negative terminal, the terminology *current* or *current flow* should be used. (Actually, current flow is a grammatical redundancy because "current" already suggests the process of flowing. However, current flow has gained such popular usage that it is considered acceptable by most editors.) Although the concept of current flowing from the positive terminal of the active device and returning to the negative terminal is an obsolete one as far as electrical theory is concerned, it has gained such a strong foothold in technology over the years that it is often referred to as the "conventional" direction of current.

The direction of the electron flow is actually the *opposite* of conventional current flow. The moving electrons, which constitute the electrical current in metallic conductors, leave the negative terminal of the source and return to the positive terminal. To describe such a situation, the terms *electron flow* or *electron current* flow are used. In this book, I use the notion of *electron flow* when I am considering the directional property of current. However, wherever feasible, a technique will be used that neatly circumvents the confusions discussed on previous page.

The simplest dynamo

The arrangement shown in Fig. 1-18 implements the principle used in the first practical electric motor and generator. At first inspection, the terms *homopolar* or *unipolar* appear to be misnomers. Two magnetic poles are, indeed, present, and any allusion to the elusive magnetic monopole was surely not intended by the originators of these names. (The machines are also referred to as acyclic types.) The basic operational difference from conventional motors and generators is that the *active conductor* interacts with the magnetic field only once per revolution. Note that whichever segment of the disk happens to be within the field is inherently the active conductor. In conventional machines of even the simplest type, a given armature conductor interacts with the field twice per revolution, and in so doing, it either generates or requires an alternating current. In contrast, the homopolar is a true dc machine. It needs no commutator to convert dc to ac, or vice versa.

This method of developing motor and generator action is often thought of as a primitive stepping stone to techniques representing greater technical sophistication. However, the homopolar machine represents a very practical approach for certain applications, and has been built in sizes ranging up to several thousand kilowatts. It is basically a high-current low-voltage machine, but higher voltages have been handled by extending the construction to accommodate multiple disks. This machine is destined to enjoy revived consideration because modern material technology, devices, and techniques make it again amenable to a hope that has never died—the elimination of the commutator. Current is conducted to and from the disk by slip-ring techniques. This has proved an obstacle in high-current applications; however, many designers would rather contend with this problem than that of clean and reliable commutation.

To illustrate what is implied by "modem methods," consider the following: homopolar motors and generators have been made in the several-hundred to several-thousand horsepower range and make use of superconducting field windings. Such field excitation involves fluxes of 60,000 gauss, and even stronger fields will be feasible. A shipboard propulsion motor of this type tends to be about one half the phys-

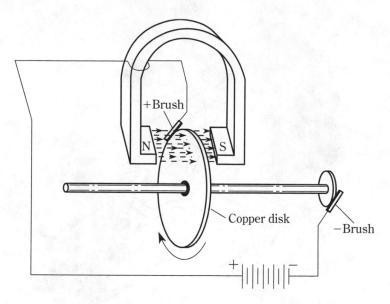

1-18 The homopolar machine devised by Michael Faraday.

ical size of ordinary motors. In one design, contact to the disk is made by means of a liquid metallic alloy that has more than twenty-five times the current-carrying capacity of brushes. The success thus far indicated suggests the increase of capacity to the 30,000-horsepower level. When combined with solid-state control techniques, these grown-up Faraday disks can be expected to become competitive, rather than archaic.

The commutator and the field poles in dc motors

The simplest arrangement of a conventional cylindrically-shaped dc motor is shown in Fig. 1-19. The commutator reverses the polarity of the applied dc as the single-loop armature turns past alternate north and south magnetic poles. This ensures that torque continuously acts in the *same* direction. If slip rings were used, rather than a split-segment commutator, continuous rotation would *not* be developed from the dc source (as in the arrangement of Fig. 1-6). Although toy motors are made according to the drawing of Fig. 1-19, this constructional format, though educational, exhibits at least two shortcomings, which mitigate against its consideration for practical designs.

When the single-loop armature rotates into the neutral plane, torque-producing action is lost. See Fig. 1-20. The commutator position is such that current through the loop is switched off. Even if such current interruption could be overcome, or greatly reduced, the relative orientation of armature and field magnetism is not proper for turning force to be exerted on the loop. What actually happens is that mechanical inertia carries the rotation of the armature loop through the neutral zone and continuous rotation *does* take place. Nevertheless, an undesirable situation develops because the motor will not self start if the armature loop is positioned in the

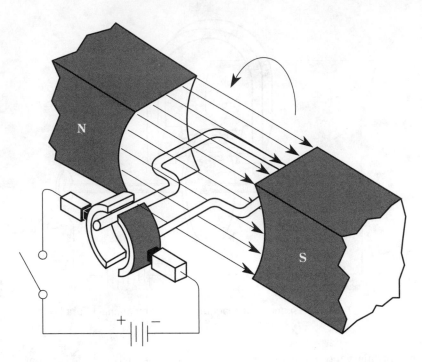

1-19 Simple dc motor showing function of the commutator. The commutator and brushes constitute a rotary switch mechanically synchronized to the position of the armature loop. Continuous rotation takes place because of reversal of armature current at appropriate times.

neutral plane. Also, with a load, such a motor will have a *pulsating* characteristic suggestive of a one or two cylinder internal combustion engine.

A partial remedy is shown in Fig. 1-21. Depicted is the addition of a second loop, together with a second pair of commutator segments. The nice thing about this format is that the motor will *always* be self-starting even if one loop is in the neutral plane. Then the *other* loop will be in a suitable field to produce starting torque. Practical motors can have many armature loops, together with correspondingly numerous commutator segments.

In such practical motors, the many armature loops would be connected in series so that current would flow through more than just a single loop at a time. In this way, torque can be produced by *many* loops even though there are only two brushes. In such a situation, the analogy is close to that of a multicylinder engine in which smooth, nearly steady torque is developed throughout each revolution. Thus, the complete remedy to the shortcomings of the simple single-loop motor is the use of many series-connected loops.

Now, a word about the field structure is in order. Books tend to use two-pole machines to illustrate many things about motors. This is because it is simpler to draw the two-pole format, and it is less likely to confuse the reader. However, actual motors can have four, six, eight, or more poles. Multipole machines provide even more smoothening of torque. Also, multipole designs enable desirable torque levels to be had at *slower* speeds. This can dispense with gear boxes or belt systems in many applications.

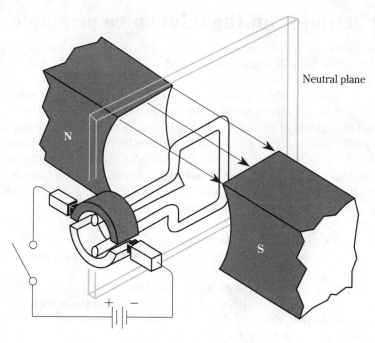

1-20 The neutral plane position of the armature loop.

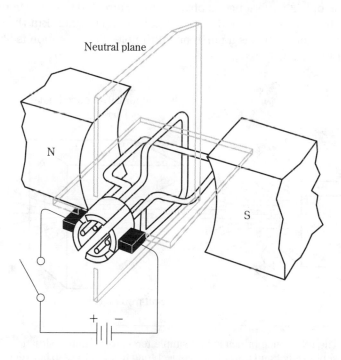

1-21 The first step toward a multiconductor armature. Adding loops together with pairs of commutator segments makes the dc motor self-starting at any shaft position, and smoothes the torque characteristic.

Motor action from the reluctance principle

The basic requisite for motor action is the interaction of forces, which can project their influence through a space. The way in which such interaction is brought about gives rise to the many different kinds of practical motors. There is yet at least one other kind of interaction of magnetic fields that assumes importance in the design of actual motors.

Suppose a child is playing with two bar magnets; if perceptive, the child might observe the possibilities of both attraction and repulsion, and might even notice the consistency with which these effects can be demonstrated. Having attained this degree of awareness, only a small step would be needed to make another intellectual advance; the child might notice the somewhat different behavior between *one* bar magnet and various soft-iron objects. Such objects, like the *other* bar magnet, are moved, but always *towards* the influencing magnet, never away from it.

No permanent magnetic-poles are induced in the soft-iron objects. The *temporary* magnetic poles of these objects always favor attraction. This is best explained by the concept of reluctance, in which the formation of temporary poles in soft iron is but the secondary result of *another* phenomenon.

Magnetic "lines" of force tend to form closed loops. By nature, these magnetic force-fields seek the *easiest* available paths in the closure of these loops. Soft iron affords easier passage than does air. Soft iron allows *denser* packing of the lines than does air. It is this increased flux density in the iron that accords it preferential treatment by the field. Because the field attempts to maximize this preferential treatment, the soft-iron object experiences a physical force directed to the region of highest flux density in the field—that is to the closest pole of the magnet. But the soft-iron object doesn't care whether this is a north or south pole. This situation is illustrated in the drawing of Fig. 1-22.

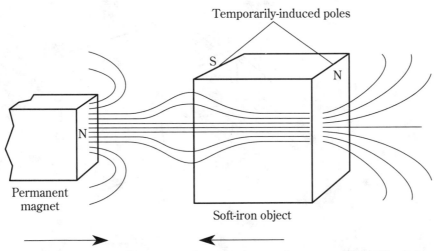

1-22 The reluctance principle in a simple force-at-a-distance situation. The permanent magnet and the soft-iron object tend to close the gap between them. Physical movement occurs so as to minimize the reluctance in the path of the magnetic lines of forces. This is tantamount to maximizing the flux density in the soft-iron object.

The term "reluctance" in the magnetic circuit is analogous to "resistance" in the electrical circuit if the terms are defined with care. Low reluctance in a magnetic material allows high flux density with minimal magneto-motive force. Low resistance in the electric circuit allows high charge density with minimal electromotive force. A little contemplation shows that electric motors can have three types of rotors as follows: current carrying coils, permanent magnet, and soft-iron rotors. More than one type can be present in order to secure special effects, such as starting torque or oscillation damping.

Electric motors at the microscopic level

The unbelievably tiny motor sketched in Fig. 1-23 is fabricated by an etching technique like that used to produce semiconductor devices. The rotor spins freely on its bearing and is just 100 micrometers in diameter. The gap between rotor and stator poles is 2 micrometers. When appropriate, sequenced 80-volt pulses are applied to the twelve stator poles, rotational speeds up to 2500 RPM have been observed. Because the operation is dependent on the programmed excitation from a control IC, the performance is suggestive of a synchronous motor. Not shown, is an electric shield underneath the structure. Notice that there are twelve-stator poles, but only eight-rotor poles.

This device has been called a variable capacitance motor. The coulomb forces that generate the torque identify it as a true electric motor. (Conventional motors that we are familiar with are actually magnetic or electromagnetic motors). At this point, it is interesting to recall the macro-sized electrostatic voltmeter, which also

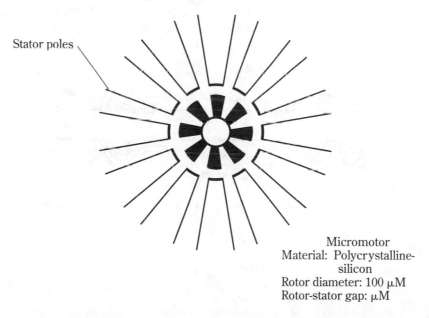

Stator poles

Micromotor
Material: Polycrystalline-
silicon
Rotor diameter: 100 μM
Rotor-stator gap: μM

1-23 Mechanical motion in miniature—the micromotor. Electrostatic force turns the eight-pole rotor at speeds up to 2500 RPM when the stator poles are appropriately sequenced with 80-volt pulses.

translates coulomb force into mechanical motion, although not in the form of continuous rotation.

It is not easy to postulate practical applications for this diminutive device. The fact of its accomplishment is, in itself, rewarding to the creative spirit. No doubt, however, unique uses will be found in medicine, instrumentation, and in various surprising ways that elude our logic. In science, as well as in invention, sometimes the discovery or innovation occurs first, to be followed by stumbling upon practical applications. A situation that quickly comes to mind is the identification of helium in the spectra of the sun before the gas was found in the earth; here an academic achievement with little apparent practical value was later put to many useful purposes.

Unfortunately, the basic motor action principles of this micromotor cannot be scaled up to produce a macro-sized motor with useful torque. The frustrating difficulty is that impractically high kilovoltages would be required. Yet, as hinted previously, it is not inconceivable that a way to harness coulomb forces for useful mechanical work might some day be developed. One can, for example, entertain such ideas as housing a high-voltage electrostatic motor in an evacuated vessel, with mechanical motion coupled to an external shaft via permanent magnets.

The ball bearing motor

An interesting scheme for the conversion of electrical energy to mechanical energy is the so-called ball bearing motor depicted in Fig. 1-24. It is, indeed, just what its name suggests—ball bearing racers mounted on an axle. Rotation takes place under the influence of a heavy dc or ac current from a very low impedance source. Speeds of several thousand RPM have been observed, but there is little reliable correlation among performance parameters, with the exception that available torque increases with current. Self starting occasionally occurs, but it is not predictable or reliable; an initial spin in either direction is usually needed.

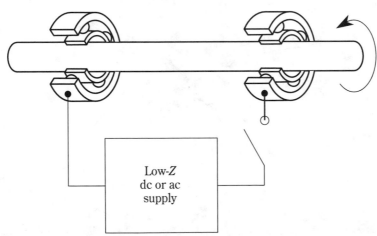

Low-Z
dc or ac
supply

1-24 Motor action is an arrangement comprised of an axle and ball bearings. This is apparently a thermoelectric rather than an electromagnetic phenomenon. Extended theoretical and empirical work will be required to make such a device practical.

Such a motor is intriguing to the experimentally inclined, but so far its low efficiency and erratic nature preclude it from practical applications. Its operation evidently is not premised on electromagnetic interactions. Rather, it appears to be an electro-thermal device. High localized heat is developed in the region of contact between the balls and their racers. This, in turn, tends to distort the spherical shape of the balls and is accompanied by strong physical forces. The affected balls attempt to reduce such "unbearable" stress by repositioning themselves. In so doing, these movements can be translated into continuous rotation between the inner and outer racers.

It is conceivable that a more refined principle of operation followed by focused developmental work could lead to a practical device with a specialized technology of its own. The compelling attributes of such an electric motor might well be its inherent simplicity and its ultimate high reliability and low maintenance. One problem that will have to be dealt with is its behavior as a near short circuit because there is no counter EMF as with conventional electromagnetic motors.

2
The classic dc motors

THE BASIC PRECEPTS OF MOTOR ACTION PRESENTED IN CHAPTER 1 NOW LEAD TO consideration of more specific matters. In the interest of ultimately dealing with electronic controls, it serves our objective to discuss the operating characteristics of motors and generators. In so doing, I will touch on the "old" (and often prevailing) methods of controlling rotating machines. It is hoped that the discussions of *operational behavior* and *control characteristics* will set the stage for the final chapters in which the actual electronic control techniques are discussed. The rationale for such a format is that, unlike the staid, rarely changing methods used up to now to control electrical machines, the methods of electronic control technology are dynamic, subject to constant improvement, and ever increasing sophistication, and are obligingly responsive to one's creative prowess.

This chapter deals with the important features of the classical direct current machines. What is implied by the reference to "classic" motors and generators? It has happened that over a span of many years, certain types of machines have become associated with unique operating behavior. Although it might be that no systematic classification was ever formally devised, it certainly is a fact of life that one has evolved. From the point of view of applying electronic controls, this is very much in our favor. The first thing to know when implementing electronic control devices are the "natural," or noncontrolled, characteristics of the motor or generator. The second thing is how was control accomplished prior to the advent of solid-state devices.

An interesting fact relevant to "classic" dc machines is that they are actually *ac devices*. The contradictory nature of such a statement stems from the fact that the rotating armatures of these machines carry alternating current, not direct current! That is why the *commutator* can be found on both motors and generators. In what is ordinarily termed a dc motor, the commutator changes the applied dc to ac for actual application to the armature conductors, while in the dc generator, the very same commutator is used to rectify the generated ac so that an essentially direct current is delivered to the load. An immediate consequence of such a revelation is that the armatures of these machines must be fabricated of *laminated* sheets, as are other ac devices such as transformers. Such construction greatly diminishes the induction of dissipative eddy currents within the iron of the rotating armature. If only dc were involved, such a scheme would be unnecessary.

Enhanced motor action from the use of iron or other ferromagnetic materials

Another approach, generally more practical, to motor action is similar to that described for the electrodynamic wattmeter and the commutator-type watt-hour meter. However, the current-carrying conductors are exposed to magnetic flux from iron-core electromagnets or from permanent magnets. Because of the much more powerful magnetic fields, significantly higher torques are developed. And, as will be shown, the presence of iron brings about important operational differences in practical motors as well. In Fig. 2-1A, a current-carrying conductor is situated in the magnetic field produced by either a permanent magnet or an iron-core electromagnet. This conductor experiences an upward force. More specifically, the force is also directed at right angles to the magnetic flux supplied by the magnet. A downward, or oppositely directed, force would be experienced if the current through the conductor were reversed, *or* if the north and south poles of the magnet were reversed. (If both the current and the poles were changed, the upward thrust of the conductor would remain unchanged.)

The magnitude of the force developed on such a current-carrying conductor is proportional to: the product of the magnetic field strength from the magnet; the amount of current carried by the conductor; and the length of the conductor exposed in this manner to the magnetic field. In most practical devices, multiple conductors are involved rather than just a single conductor. The windings on motor armatures, for example, are intended to accomplish such a situation. The mechanical force produced must then also be proportional to the number of current-carrying conductors, as well as to the three factors just mentioned.

Notice in Fig. 2-1 that the composite field in the immediate vicinity of the current-carrying conductor is *distorted*. Indeed, without such field distortion, there can be no force developed on the conductor. For example, if no current flows through the conductor, no circular magnetic field will surround the conductor, and there will be no distortion of the main field. In turn, no force will be experienced by the conductor. These apparently simple facts have important ramification in electric motors. On the one hand, the more field distortion, the better. But excessive deflection of the main field also brings about certain operating difficulties.

The D'Arsonval meter movement shown in Fig. 2-2 is an example of motor action with the use of iron. In this device, the stationary iron core deflects the magnetic field as shown so that the scale divisions can be uniformly spaced. Although this particular scheme is not used in motors and generators, the manipulation of field direction in rotating machines is very important with regard to obtaining proper commutation.

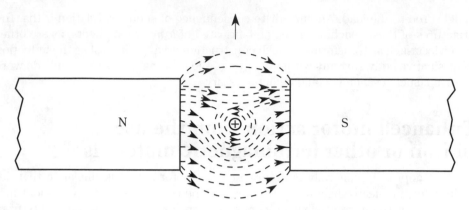

A. Motor action: current-carrying conductor develops upward force.

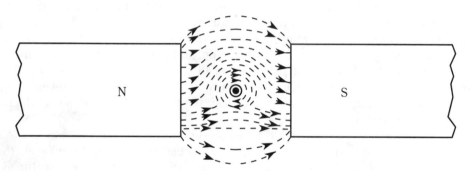

B. Generator action: An EMF with current-supplying capability is developed in conductor, which is moved upward through magnetic field by an external force.

2-1 Current-carrying conductors in fields produced by ferromagnetic material.

Motor and generator action

Returning to Fig. 2-1, the cross signifies electron current directed *into* the page; the dot signifies electron current directed out of the page. In Fig. 2-1B, no external source of voltage is employed to drive current through the conductor, and the upward motion of the conductor, instead of being *developed*, is *supplied*. This is basic generator action. (Of course, the conductor must be part of a complete circuit for actual current to flow.) Internally, the field distortion has changed. At this point, the change has been such that it opposes the direction of physical motion supplied to the conductor. (This is a manifestation of Lenz's law.)

The significance of Fig. 2-1 is not so much that the motor and generator actions are reciprocal functions—you supply electrical energy and derive mechanical energy, or conversely, you supply mechanical energy and induce an EMF with current-

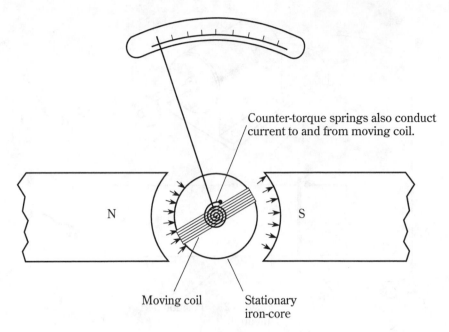

Counter-torque springs also conduct
current to and from moving coil.

N

S

Moving coil

Stationary
iron-core

2-2 The D'Arsonval movement is an example of use of ferromagnetic materials.

delivering capability. The significance, at least from the standpoint of control techniques, is that these two functions occur together in any one machine, whether it is
called a *motor* or a *generator*. The voltage induced in the moving conductors of a
machine operating as a motor is called the *counter EMF*. For a machine to run as a
motor, the voltage applied to its terminals must be greater than the induced voltage,
or counter EMF, from its own generator action.

In a machine that is operating as a generator, the internal motor action opposes
the torque supplied to the machine. In order for the machine to run as a generator,
the applied torque must overcome the internally produced torque. Referring now to
Fig. 2 3, try a simple experiment that will clarify the simultaneous motor/generator
action described earlier. First, let the battery and the voltmeter be connected to the
dc motor (POSITION 1). Then disconnect the battery, allowing the motor to coast to
standstill (POSITION 2). When the disconnect is first made, the voltmeter reading
becomes only slightly less than when the battery is supplying motor current. The
machine now is functioning as a *generator*, just as it did prior to the disconnect
when its primary function was that of a *motor*.

In the preceding experiment, if a load is placed across the machine's terminals
while it is still coasting (POSITION 3), it very quickly slows down and comes to a
halt. This demonstrates that it is internally developing *motor action* while functioning *primarily* as a generator. This motor action obviously opposes the direction of
rotation while the machine is behaving as a generator to deliver electrical energy to
the load. As can be anticipated, these characteristics will be found intimately related
to control techniques.

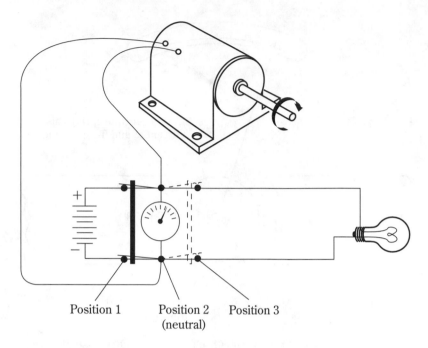

Position 1   Position 2   Position 3
             (neutral)

2-3 Demonstration of simultaneous motor and generator action in same dc machines.

A disadvantage of the commutator

At an appropriate instant during the rotation of a dc motor, its commutator must reverse the current through those armature conductors, which having left the influence of a field pole, are approaching that of an alternate pole. In this way, unidirectional torque is produced, leading to continuous rotation. But, the physical fact that a brush can contact more than one commutator segment is not a trivial matter because an armature loop can thus be shorted out. Obviously, the short endures for only a tiny fraction of the time required for one revolution of the commutator. However, if the shorted armature loop has a difference of potential across its ends, severe sparking can occur at the interface of the brush and commutator segments. At worst, such sparking produces burning and pitting of the commutator; at best, it results in erratic brush contact and the generation of needless RFI (radio frequency interference). Because there is no way to avoid the shorting of adjacent commutator segments, it is necessary to ensure that no voltage is induced in the commutator loop undergoing the momentary short.

The shorted condition is progressively experienced by all of the loops as their commutator segments rotate under the brushes. The action is similar for both brushes. Whether the machine is operating as a motor or as a generator, voltage will be induced in the rotating armature loops when they "cut" magnetic lines of force in the main field. If the short is occurring at the instant the active conductors in the armature loop are moving parallel to the field, no generator action takes place, apparently making it "timely" to accommodate the momentary short at the commutator segments. The vertical axis occupied by the shorted armature loop is known as the geometric neutral plane. It would appear that, in principle, such a technique should result in clean, sparkless commutation. Unfortunately, in actual machines, the situation is somewhat more involved.

The ensuing discussion show that the geometric neutral plane is more important than our description might imply. For more than one reason, the geometric neutral plane is not necessarily the place where the commutator can be allowed to undergo its shorting action. In any event, keep in mind that it is the current-reversing action of the commutator/brush system that is desirable—the shorting action is not. This situation, incidentally, has spurred the development of commutatorless machines in which the current-reversing process is accomplished with solid-state devices.

Armature reaction

After you grasp the fact that *simultaneous* motor and generator actions occur in dc machines, it becomes relevant to consider factors affecting *commutation*. The illustrations of Fig. 2-4 depict the field that an armature is exposed to under three conditions. In Fig. 2-4A, there is no armature current. If the machine is a motor, its armature is not connected to the dc power source, but its field is energized. (Actually, the field can be produced by either permanent magnets or by electromagnets.) If the machine is a generator, it is not being driven. Therefore, like the motor, it has no current in its armature conductors. The situation is then identical in both machines. In both the motor and the generator, the geometric plane of the field coincides with that of the field magnets themselves. More significantly, the armature conductors move *parallel* to the field when they are at the tops and bottoms of their rotations. At these positions, the conductors are said to be in the electrically neutral plane, and they can neither have voltages induced in them nor contribute torque. In this idealized situation, the geometrical and electrical neutral planes are identical.

Contrary to the described idealizations, the actual field patterns of motors and generators are shown in Figs. 2-4B and 2-4C, respectively. In Fig. 2-4B, the motor is driving a mechanical load. In Fig. 2-4C, the generator is delivering power to an electrical load. In both cases, the field is distorted because of the magnetism associated with the current-carrying armature. For the motor, the electrical neutral plane has been shifted backward from the direction of rotation; for the generator, the shift has been in the direction of rotation. Appropriately, this effect is termed *armature reaction*. It is a cumulative effect of the simple one-conductor distortions shown in Fig. 2-1. Because of armature reaction, the timing of the motor and generator action—and particularly of the *zero* point in voltage induction and torque production—is altered. This plays havoc with commutation. (The drawings in Fig. 2-4 involve simplifications. For example, in Fig. 2-4A the field flux does not actually pass through the armature iron and the steel shaft as if they were air. However, the three illustrations provide insight into the phenomenon—the deflection of main field flux in the vicinity of the armature conductors.)

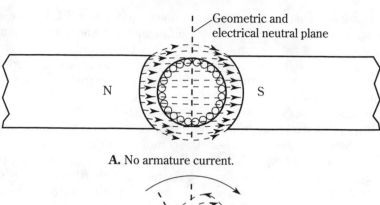

A. No armature current.

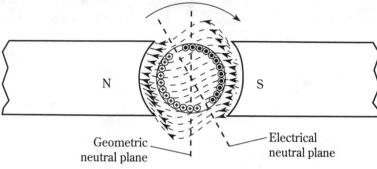

B. Armature current in a motor.

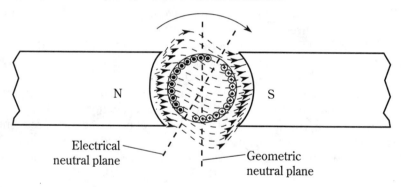

C. Armature current in a generator.

2-4 Field distortion produced by current-carrying conductors in dc machines.

Armature self-inductance

It's clear that considerably more is involved in commutation than the mere switching or reversing of current. And armature reaction—that is, the distortion of the main field by the magnetic field associated with the armature conductors—is not the only difficulty encountered. Another difficulty is the *self-inductance* of the armature. Even if an armature loop is shorted while progressing through main field flux that is parallel to its motion, considerable sparking can occur. It is true that no generator action is then taking place. But, because of the self-inductance of the armature, the current that had previously circulated in the shorted armature loop cannot be instantly extinguished. Thus, full advantage cannot be taken of the fact that the shorted armature loop does not cut field flux. In other words, what appears to be a neutral plane is not, as far as clean commutation is concerned.

The effect described is often lumped under the general heading of "armature reaction." However, I have clearly delineated two separate phenomena. Armature reaction has to do with the bending or distortion of the main field flux. Armature self-inductance has to do with the fact that armature current requires time to die down to zero after field flux is no longer being cut. The reason the two phenomena are frequently lumped under the heading "armature reaction" is that armature self-inductance behaves in some respects as if it were, indeed, additional armature reaction.

A straightforward remedy for armature reaction and armature self-inductance is an adjustable brush assembly. In a motor, the brushes must be shifted *backwards* from the direction of rotation in order to achieve clean commutation. The reverse is true in generators. In both cases, the shift of the brushes away from the geometric neutral plane is more than sufficient to avoid the effects of armature reaction alone. Additional brush displacement is required to allow time for the current delayed by armature self-inductance to approach zero. When clean commutation is thereby attained, the brush axis can be said to lie in the true electrical neutral plane of the main field.

Canceling armature reaction with compensating windings

The adjustable brush technique obviously has manufacturing and economic shortcomings. Its greatest disadvantage, however, is that a given brush adjustment yields optimum performance for only one armature current. This might not be considered overwhelmingly unfavorable for some applications, but in general, a remedy that is independent of current is desirable. Experiments have been conducted using electronically actuated servo controls to automatically maintain the brushes at optimum positions for clean commutation, despite wide variations in armature current. Although it is too early to judge this approach, it appears suitable primarily for large machines. However, all things considered, it would be better to dispense with such mechanical motion.

An accepted technique for accomplishing near cancellation of armature reaction is shown in Fig. 2-5, where so-called compensating windings are placed in slotted faces of the main pole pieces. In this diagram, as well as in most of the dc motor and generator illustrations in this book, two-pole structures are depicted. This is done for the sake of simplicity; practical machines often have two or more pairs of poles. Also, the main field poles are usually shown as being simple permanent magnets. Although the main field poles can be permanent magnets, they are more often electromagnets. And, unless stated otherwise, the windings on the main poles might be a relatively few turns of heavy wire that carry the full armature current, or they might be many turns of fine wire that are connected across the terminals of the armature. These two cases correspond to series- and shunt-type machines, respectively. Although such machines have widely differing characteristics, a general discussion of

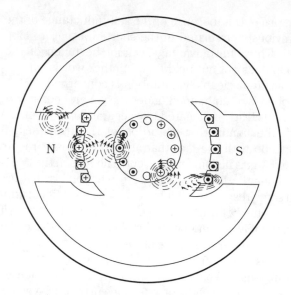

A. The effect of the armature conductor on the main field is opposed by the adjacent field of the compensating winding.

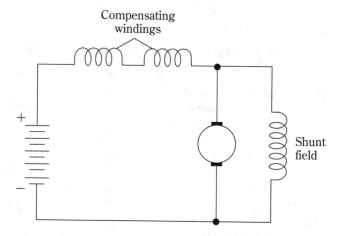

B. Circuit connection for the compensating winding.

2-5 The use of compensating windings for canceling armature reaction.

commutation, armature reaction, and armature self-inductance is made easier if it is assumed that the way in which the main field is produced is not of primary importance. Thus, it would be distracting to show main field windings and their connections in Fig. 2-5A. Basically, it is desirable to focus attention on the way in which the poleface conductors tend to cancel the magnetic fields that the armature conductors would otherwise contribute to the air gap. In Fig. 2-5B, the motor is shown with a shunt field to emphasize that the compensating windings are an addition to an otherwise "normal" motor.

The use of interpoles to improve commutation

Another method of improving commutation is to induce a reverse-polarity current in the armature loop as it approaches commutation. The rationale here is that of "bucking out" whatever current would otherwise flow in this loop when it underwent a short. This is brought about by the use of *interpoles*, or *commutating poles*, which are relatively small poles positioned halfway between main poles. Figure 2-6 depicts the basic idea for a simple motor structure.

It is often stated that the interpoles are provided to negate the effects of armature reaction. Although such a statement is essentially true, the interpoles are primarily intended to combat the effects of self-induction and mutual induction in the armature conductors. This is especially so when both compensating windings and interpoles are used. (Mutual induction pertains to voltage induced in an armature loop from adjacent loops. Like self-induction, it causes a voltage to exist across the ends of a loop even though no generator action is occurring with respect to the main field.)

It so happens that when interpoles alone are used, they can help overcome the effects of armature reaction. That is, they can cause a reverse current in a shorted armature loop strong enough to buck out not only currents that would flow because of self-inductance and mutual inductance, but also those currents due to generator action in a distorted main field. In this sense, the interpoles provide remedial action to armature reaction and thereby improve commutation. In small machines, the brushes can be permanently positioned in the geometrical-neutral plane when interpoles are used. The interpoles, like compensating windings, are connected in series with the armature, as shown in Fig. 2-6B, and therefore are effective over a wide range of load conditions. Under severe overload, however, the interpoles tend to saturate, which diminishes their effectiveness and leads to poor commutation.

As already suggested, the best overall results are obtained from the use of *both* compensating windings and interpoles. Often, this is economically justifiable only in larger machines.

Additional techniques for improving commutation

Frequently, dc machines are encountered in which the brush axis is in the geometrical plane and neither compensating windings nor interpoles are utilized. However, there might be more than meets the eye in the design of the machine. For example, Fig. 2-7 illustrates three techniques that can be quite effective in negating the commutation-degrading effects of armature reaction, self-inductance, and mutual inductance associated with the armature conductors.

Figure 2-7A depicts an armature loop being shorted by a brush bearing on the two commutator segments to which the ends of the loop are connected. The brush

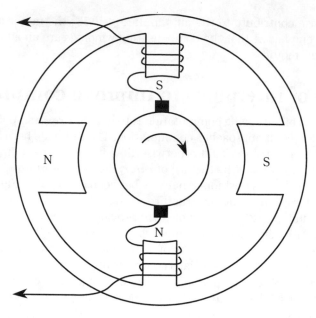

A. Arrangement of interpoles in a simple motor.

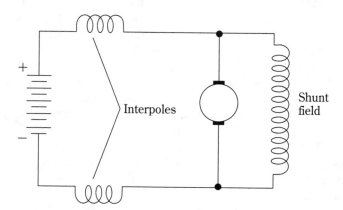

B. Circuit connection for the interpoles.

2-6 The use of interpoles for improving commutation.

is processed so that its resistivity is reasonably low in the brush plane that accommodates current to, or from, the armature. In the illustration this is the vertical plane. But, the resistivity *crosswise* through the brush is relatively high—that is, in the horizontal plane along which the short circuit current must flow, high resistivity is encountered.

The scheme shown in Fig. 2-7B uses a pole structure shaped so that the magnetic reluctance in the regions of the pole tips is greatly increased. This can be brought about by the use of a nonconcentric curvature of the pole face so that the radial dis-

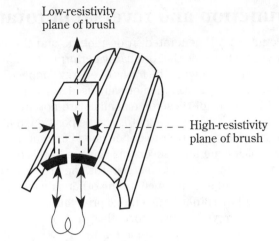

Low-resistivity
plane of brush

High-resistivity
plane of brush

A. Specially processed brush displays high resistivity to short-circuit current.

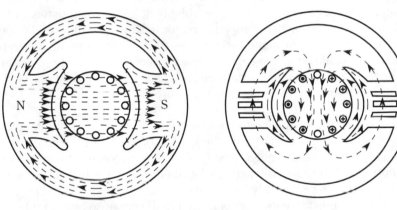

B. Weak flux in vicinity of pole tips
opposes main field distortion as a result
of armature reaction.

C. Slots in main pole pieces offers greater
magnetic reluctance for cross flux from
armature than for main field flux.

2-7 Additional techniques for improving commutation.

tance between the armature and the pole face is greater near the tips than in the central region. In the illustration, field flux is shown but no current is flowing in the armature conductors. Notice that the field density is relatively sparse in the regions of the pole tips. This condition is opposite to that occurring with conventionally shaped field poles and with armature current flowing. Therefore, the "pinched" field configuration shown to produce when current is flowing in the armature.

A somewhat more sophisticated method of combating armature reaction is shown in Fig. 2-7C. Here, the pole piece is slotted to offer high magnetic reluctance to the cross field associated with the current-carrying armature. The armature flux, therefore, has diminished strength with which to distort the main field. In this example, it is expedient to show the armature flux instead of the main field flux.

Machine function and reversal of rotation

Because of the counter EMF generated in dc motors, and the countertorque produced in dc generators, a machine that ordinarily performs the function of a motor becomes a *generator* by interchanging the forms of energy supplied and extracted. The reverse interchange converts a machine normally used as a generator into a motor.

The reciprocal functions of these dc machines are not always so simply implemented. For example, the brushes of a motor can be positioned at some angular distance from geometric neutral in order to obtain good commutation. If the machine is pressed into service as a generator, severe sparking is likely to occur. For the same armature current in a machine used as a generator, the proper brush position is now approximately the same angular distance on the other side of the geometric neutral axis. Also, brushes that have other than a radial orientation with respect to the commutator do not always provide satisfactory electrical or mechanical performance when the rotation of the armature is reversed.

A common trouble encountered in reversing motors stems from ignoring the series field in *compound motors*. If, for example, such a motor is reversed by transposing the connections to the armature alone, the operating characteristics will be changed from those of a cumulative compound motor to those of a differential compound motor, or vice-versa. The same logic applies to the compound dc generator when the polarity is changed in the same manner. In both compound motors and generators, the leads common to the armature and the series field must not be disconnected. The two remaining leads are then transposed in order to reverse rotation in the motor, or reverse polarity in the generator.

The same reasoning also applies to compensating windings and to interpoles. However, the situation is less likely to involve trouble or confusion with such connections because they are usually not as accessible. For example, one armature lead of a shunt motor with interpoles will actually be the commutator brush. The other "armature" lead, however, will usually be one end of the interpole coils. (The other end of the interpole coils is permanently connected to the opposite brush.) Whether the machine functions as a motor or as a generator, and regardless of its rotation, such poles or windings will always be properly connected.

The simplest case of all pertains to having no interpoles, compensating windings, or compounding windings, and in which the brush axis coincides with geometric neutral. With such machines, equally acceptable commutation can be obtained regardless of direction or function.

Speed behavior of dc shunt motors

The speed control characteristics of the dc shunt motor are particularly important because they are often the basis for comparison with other types of motors. Of special interest is the speed behavior with respect to variable field current. Does greater field strength make the speed increase or decrease? This is hardly a trivial question; unless you are already knowledgeable in this matter, it's easy to get confused. For example, the three control situations depicted in Fig. 2-8 all achieve speed variation by controlling the field current. Yet, the individual behavior of each appears to con-

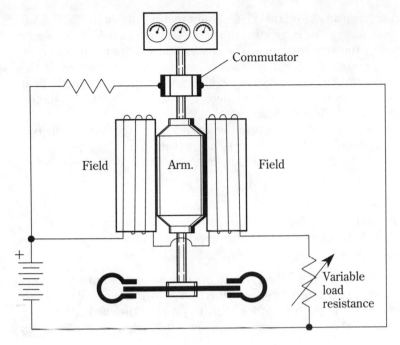

A. A dc watt-hour meter.

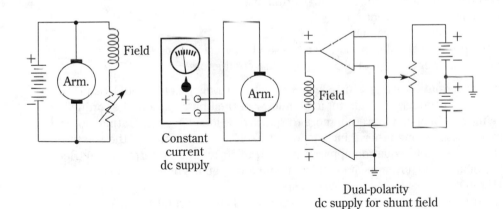

B. A conventional dc shunt motor.

C. A servo motor. The field is operated from an amplifier arrangement that allows for polarity reversal in order that the motor can be run in either direction.

2-8 Field control of dc shunt motors.

tradict rather than confirm any basic rule. The "shunt motor" shown in Fig. 2-8A is actually a dc watt-hour meter. The more current passed through its field, the *greater* the speed developed by this small machine. The conventional shunt motor shown in Fig. 2-8B behaves in opposite fashion; the greater the field current, the *slower* the speed. Finally, the motor of Fig. 2-8C is again a shunt motor, although it might bear the specialized nomenclature of "servo motor." Here, as with the dc watt-hour meter, increased field current results in *higher* speed.

In order to gain insight into these apparent inconsistencies, try to reconcile two equations, both of which are valid but appear to suggest divergent interpretations. The generalized formula for the torque, T, of any dc motor can be stated as follows:

$$T = k \, \Phi \, I_a$$

where:
   k represents the number of poles, armature paths, and conductors,
   Φ is the flux per pole that links the armature conductors, and
   I_a is the armature current.

Don't be concerned with units; rather, try to pin down the qualitative reasons for the divergent speed characteristics described on previous page. Use these equations to facilitate this objective, rather than to provide the background for numerical solutions. The second equation, which gives the speed of dc motors, is stated as follows:

$$S = \frac{V_a - (I_a R_a)}{k \, \Phi}$$

where,
   S represents speed,
   V_a is the voltage across the armature terminals,
   I_a is the armature current,
   R_a is the armature resistance,
   k is the number of poles, armature paths, and conductors, and
   Φ is the flux per pole that links the armature conductors.

The first equation leads you to believe that a stronger field should result in greater speed. Although speed is not a parameter of the first equation, the torque is a "prime mover" of speed. Surely, if more torque is developed in an operating motor, the torque must, among other things, manifest itself as acceleration of the rotating armature. The second equation apparently contradicts such an interpretation. Because the quantity Φ now appears in the denominator, you might expect an increase in field strength to produce a *decrease* in speed.

Unfortunately, neither of these equations reveals an important characteristic of dc shunt machines. The armature current and field flux in both equations are not independent quantities. The numerator of the second equation is actually the counter EMF and can significantly affect armature current—a torque-producing factor. The counter EMF, in turn, is governed by Φ.

In conventional motors, a small change in field strength produces a much greater percentage of change in armature current. For example, if Φ is increased in the first equation, I_a will be reduced to the extent that the torque will actually *decrease*. Because of this, you must anticipate a *decrease* in speed. However, some care is needed in order to confirm this when you are dealing with the second equation. What neither equation tells us is that dc motors (and generators) are actually *power amplifiers*. In these machines, a small change in the power applied to the field is manifested by a *large* change in the power consumed (motor) or delivered (generator) by the armature. The change in armature power level is primarily caused by the variation in armature current.

Now justify the speed characteristics displayed in the three shunt motor situations of Fig. 2-8. The "air-core" shunt motor, or watt-hour meter of Fig. 2-8A, does not generate much counter EMF; that is, the counter EMF is low compared with voltage V_a impressed across the armature. Unlike the situation prevailing in conventional machines, the counter EMF is not a strong function of field strength Φ. Therefore, when Φ is increased, the first equation, $T = k\,\Phi\,I_a$, tells us that the torque, T, will also become greater. This being the case, you would expect the armature to accelerate toward a higher speed. Can this be confirmed with the second equation:

$$S = \frac{V_a - (I_a R_a)}{k\,\Phi}$$

For the qualitative purpose involved, inject some numbers that can be handled with easy mental arithmetic. Let $V_a = 100$ volts. And, because we know that the counter EMF is low, the armature-resistance voltage drop, $I_a R_a$, will be very close to 99 volts. Finally, assume that the product of k and Φ is 0.10 and that this enables the speed, S, to be expressed in revolutions per minute. Making the appropriate substitutions, we have:

$$S = \frac{100 - 99}{0.10} = 10 \text{ RPM}$$

Let the field strength be increased by 10 percent, and assume that this leads to a decrease in armature current of approximately one percent. You are guided in making such an assumption by the known generator action of this device—that is, by the low counter EMF produced. Otherwise, the quantitative significance of our numbers is of no great importance for our goal, which is to ascertain which way the speed changes with respect to field strength. Again, making the appropriate substitutions, we now have:

$$S = \frac{100 - 98}{0.11} = 10 \text{ RPM}$$

Thus, the speed has *increased* with increasing field strength. (Inasmuch as the interest here is not in the instrumental aspects of the basic shunt motor used in the dc watt-hour meter, the action of the shaft-attached, eddy-current disk has not been mentioned. However, its function is to impose a countertorque directly proportional to speed. This brings about the desirable operating feature of the speed of the overall device becoming proportional to the power supplied to the load.) With the discussion of this device as a prelude, a similar investigation for the conventional shunt motor of Fig. 2-8B is now conducted.

Because the conventional shunt motor has iron in its magnetic circuit, the counter EMF will tend to be close to the voltage applied across the armature terminals. However, the voltage drop across the armature will tend to remain relatively small. These facts will always hold true for a shunt motor that is operating at a constant speed with a fixed load within its ratings. Obviously, under such conditions, the motor must be simultaneously operating as a generator that produces a voltage in opposition to the impressed armature voltage. Only a small portion of the impressed voltage is actually available to force current through the armature resistance in accordance with Ohm's law. Assuming such a motor is operating with a moderate load at a constant speed, what happens to the speed when the field current, and therefore the field strength, is increased?

A casual inspection of the speed equation,

$$S = \frac{V_a - (I_a R_a)}{k\,\Phi}$$

certainly suggests a *reduction* in speed. The impressed armature voltage, V_a is fixed; k is a constant associated with the particular machine; and there is no change in the armature resistance, R_a. It would appear that the armature current, I_a, would have to adjust itself to the new conditions, but you have to know in which direction this change in armature current will occur and the relative magnitude of the change. Otherwise, you might be led to an erroneous interpretation when attempting to reconcile the torque equation with the speed equation. The torque equation, $T = k\,\Phi\,I_a$, seemingly suggests that, on the basis of field strength Φ alone, torque T increases. This would imply that the motor accelerates when Φ is made greater, which would lead to *higher* speed. The fact that the motor actually slows down when its field current (field strength) is increased imposes a dilemma.

The resolution of this paradoxical situation requires insight beyond that provided by the mere mathematical interpretations of the two equations. Actually, when Φ is increased, the resultant increase in counter EMF developed in the armature reduces the armature current taken from the dc source. The reduction in armature current I_a is, by virtue of the "amplifying action" of the machine, proportionally greater than the increase in field strength Φ. Therefore, the apparently strange fact that torque T is reduced must be accepted. Reduced torque must manifest itself in a slowing down of the motor!

The slowed motor then develops a lower counter EMF. This, in turn, allows more armature current. Ultimately, the motor is operating at a slower speed but has reestablished equilibrium with the same counter EMF and the same armature current as prevailed before the field was increased.

The speed of shunt motors is ordinarily controlled over a range of field current, and the field strength, Φ, is directly proportional to the field current. Outside of this range, there is either diminishing returns from the effect of magnetic saturation of the iron in the pole structure or operational instability and poor commutation because of armature reaction.

In practice, the speed equation is not used in an absolute sense. That is, it is not generally used for predicting motor speed "from scratch." Rather, it is used to set up a proportion from previously measured parameters obtained from an operating motor. Thus, you might have recorded V_a, I_a, R_a, the field current and the speed. If the field current is increased by ten percent, it is easy enough to deduce that the speed will be decreased by ten percent.

The motor depicted in Fig. 2-8C could be the same one just discussed in the simpler control arrangement shown in Fig. 2-8B. However, the speed behavior is altogether different—now the motor speeds up in response to higher field current. A more complex field current supply and a separate power source for the armature is shown.

The singular item responsible for this perhaps unexpected speed characteristic is the constant current supply for the armature. It is the basic nature of a constant current source that the voltage delivered to a variable load varies widely. It is this varying voltage that enables the current delivered to the load to be maintained constant. This is one of two important factors in this speed-control technique. The other factor is the influence of the applied armature voltage, V_a, in the speed equation:

$$S = \frac{V_a - (I_a R_a)}{k\,\Phi}$$

A little more consideration will show that the speed would be almost directly proportional to V_a if it alone were varied and if all other parameters remained constant. Although no voltage-control technique is shown associated with the motor armature, the use of the constant current supply varies the voltage applied to the armature. From previous discussions, you know that a small increase in field strength produces a large reduction in armature current. Inasmuch as the armature current change would have been great with respect to the change in field strength, the constant current source is forced to make a large change in armature voltage in order to maintain the armature current constant. Therefore, armature voltage becomes the governing factor in the speed equation. Its position in the speed equation enables it to influence speed very nearly on a directly proportional basis—that is, S tends to be proportional to E_a. (Actually, the field-current supply need be nothing more involved than a battery and a rheostat. The arrangement illustrated is suggestive of a servoamplifier application in which the motor can be operated in either direction.)

Before applying electronic controls to motors and generators, it is necessary to be familiar with the "natural" performance of these machines. In general, two methods of plotting motor characteristics assume importance, although additional plotting techniques are used occasionally. In one of the main methods, the various operating parameters of a motor are plotted as functions of *armature current*. Thus, you might be interested in *speed* versus armature current, or *torque* versus armature current. For these purposes, armature current is plotted on the horizontal axis because it is considered the independent variable.

This way of looking at motor performance is primarily of interest to the engineer, the student, or anybody who wants to gain insight into the basic operating mechanism of the motor. The user is not greatly concerned with armature current once the basic requisites of wire size, fuse or circuit-breaker capacity, and power source are taken care of. (Sometimes the armature-current quantity in a shunt motor might also include the *field current*. In such a case, it is erroneous to consider the total line current the same as the armature current alone. But, because the shunt motor field current tends to be a small percentage of armature current, especially when the motor is appreciably loaded, the error of such a presentation is often not of great consequence. In series motors and permanent-magnet motors, armature current and line current are identical.)

The user is more interested in the *usable quantity* available from the shaft—this quantity is the turning power, or torque, of the motor. That is why graphs of motor performance often depict speed, horsepower, efficiency, etc., as functions of a torque, rather than of armature current.

As mentioned before, other graphical formats are also used. For example, the user might be interested in the quantity of horsepower rather than torque that is delivered by the motor shaft. Horsepower comprises the two factors of speed and torque. Torque, on the other hand, does not involve speed. Indeed, one of the important things revealed by a graph plotted against torque is the turning power of the motor at *zero speed*—this is the starting torque.

Often, the same information can be revealed by different plotting methods, and the choice is then determined by convenience of interpretation. If you were considering some kind of a control for an automotive starter motor, for example, you certainly would be interested in seeing a graph depicting speed versus torque. But, you would *also* like to see torque plotted as a function of armature current. Graphs with specialized formats to emphasize the behavior of efficiency or horsepower could conceivably prove useful too. From the standpoint of control, plots of motor characteristics versus field current can be of prime importance when dealing with shunt-field and compound-field machines.

Basic characteristics of the shunt motor

The speed and torque characteristic of a dc shunt motor are shown in Fig. 2-9. Loosely speaking, the shunt motor can be called a constant-speed machine. Indeed, the effects of armature reaction and magnetic saturation often tend to make the speed curve even flatter than that illustrated in the graph of Fig. 2-9B. At first, it might seem that the performance curves of shunt motors reveal the operation of permanent-magnet motors also. However, there can be significant differences in the region of overload. Instead of the speed encountering a "breakdown" region, such as is depicted by the dashed portion of the curve in Fig. 2-9B, a permanent-magnet motor can have a linearly sloped speed characteristic right down to zero speed. This implies greater torque at low speeds. Because a motor is at zero speed just prior to starting, the permanent-magnet motor is capable of exerting higher starting torque than its equivalently rated shunt machine (in terms of horsepower).

The above described differences are largely brought about by the different conditions of armature reaction of the two types of motors. However, much depends on the selection of magnet materials and on other design factors. It does not necessarily follow that the permanent-magnet motor will feature this extraordinary behavior. Older designs often gave poor accounts of themselves in actual service because of susceptibility to demagnetization from the effects of armature reaction. Because of advanced materials technology, and sometimes because of the inclusion of compensating windings in the pole faces, this shortcoming has largely been overcome.

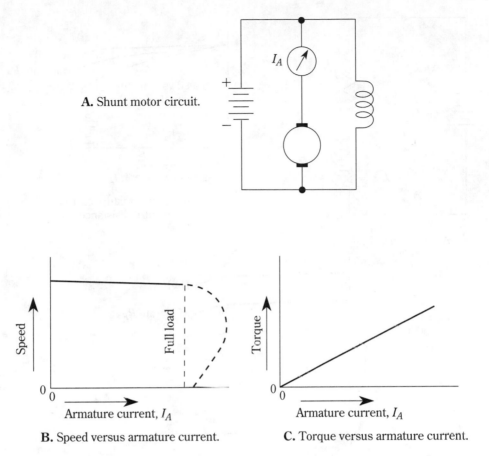

A. Shunt motor circuit.

B. Speed versus armature current.   **C.** Torque versus armature current.

2-9   Speed and torque of a shunt motor as a function of armature current.

In Fig. 2-10, the speed versus armature current behavior of the shunt motor is extended for the situation in which resistance is inserted in the *armature circuit* (R_a) and in the field circuit(R_f). The actual variation in armature current is intentionally caused by varying the mechanical load applied to the shaft. Significantly, the use of armature resistance enables lower speed ranges to be attained. In contrast, when resistance is inserted in the field circuit, higher speed ranges become available. A shortcoming of the use of armature resistance is that it deprives the motor of its near-constant speed behavior. Also, the power dissipation in this resistance presents serious problems in larger motors. From casual inspection of the graph in Fig. 2-10C, the field-resistance method of speed control displays no disadvantageous features (other than the inability to acquire lower speed ranges). However, the weakening of the field increases its vulnerability to distortion from armature reaction. Attempts to reduce speed by a factor exceeding about four to one by this method tend to cause commutation difficulties and operating instability.

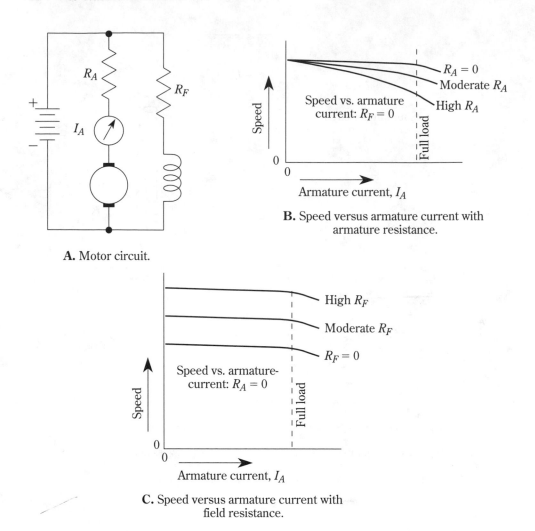

A. Motor circuit.

B. Speed versus armature current with armature resistance.

C. Speed versus armature current with field resistance.

2-10 Speed behavior of the shunt motor with armature or field resistance.

Motor speed control by shunt field current

Figure 2-11 illustrates the general situation prevailing in dc shunt motors when the speed is controlled by the field current. Interestingly, the control could be usefully extended to a speed range well below the rated speed if it were not for the magnetic saturation of the field iron. With actual machines, the control region above the rated field current is in an area of diminishing returns even if you are interested in only limited-duration applications of high field current.

   If the field current is reduced in order to operate the motor at a higher speed, the armature current corresponding to the higher speed operation is substantially the same as it was for the slower speed. The assumption is made that the torque demand of the load remains constant. Nonetheless, when the field current in a shunt motor is reduced during operation, dangerously high armature current might result, particularly in large machines.

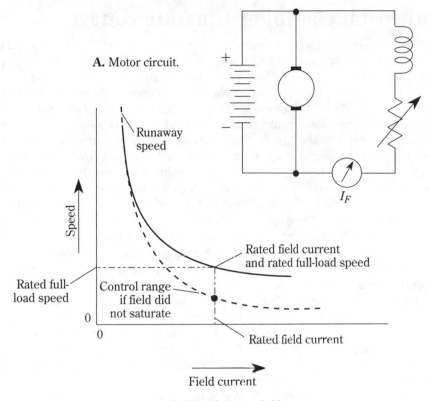

A. Motor circuit.

B. Speed versus field current.

2-11 Speed as a function of field current in a shunt motor.

The apparent contradiction in the above paragraph is resolved as follows: When the field current is first reduced, the counter EMF almost immediately follows suit. This causes a large inrush of armature current. The resultant accelerating torque will ultimately develop a high enough speed that the counter EMF will admit only enough current to accommodate the torque demand of the load. But, because of the rotational inertia of the armature, and often of the load, this corrective action cannot take place instantly. Thus, while the motor is accelerating, very high armature currents can exist.

The action described above becomes dangerous at low field currents, and particularly at zero field current. This is the "runaway" region depicted by the dotted portion of the speed curve in Fig. 2-11. The action is not qualitatively different from that described for the normal range of control by field-current variation. However, the motor now seeks a fantastically high speed to restore its operating equilibrium. Before this speed can even be approached, the armature might explode from centrifugal forces or the motor might suffer damage from the tremendous current inrush. The tendency is similar whether or not the motor is loaded; when loaded, the self-destruction by racing might take a little longer, but the damage caused by excessive current is likely to occur more quickly.

Nonrheostat control of armature voltage

Before the development of solid-state devices, specifically the power transistor and the thyristor, armature-voltage control of shunt motors was accomplished by means of rheostats or tapped resistors. The high-power dissipation involved imposed practical difficulties in implementation and would have been economically devastating in the long run, due to the waste of electrical energy. Also, the resistance in the armature circuit resulted in poorer speed regulation. It was found that, except for initial cost, the above disadvantages could be largely overcome by powering the armature of the controlled motor from the output of a generator, which in turn could have its output controlled by its own field current. The motor then would "see" a variable-voltage power source—one in which the internal resistance was always very low. A motor controlled in this manner responded by displaying excellent starting characteristics, minimal commutation difficulties, and an inordinately wide range of speed control. This control technique, known as the Ward-Leonard system, is shown in Fig. 2-12.

In Fig. 2-13A, the characteristics of an armature-voltage controlled motor are shown. The variable-voltage power supply can be a regulated power supply with low output impedance. More often, however, it is a *thyristor circuit* in which the average output voltage is varied by phase control of the firing angle. Pulse-width-modulated switching transistors can be employed in a similar manner. Such a control method can be electrically efficient and quite economical compared with older control methods. This is especially true with large machines, in which case the thyristor presently predominates. Moreover, such armature-voltage control is applicable in the same way to permanent-magnet motors.

The control method illustrated in Fig. 2-13B is similar, except that *both* armature and field voltages are controlled together. The speed and torque curves then have the general shapes shown. Starting torque is appreciably lower in this arrangement than in that of Fig. 2-13A. However, a separate field supply is not needed.

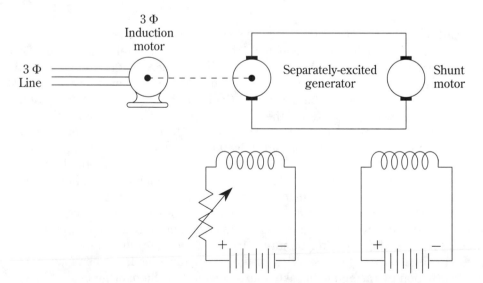

2-12 The Ward-Leonard speed control system for dc shunt motors.

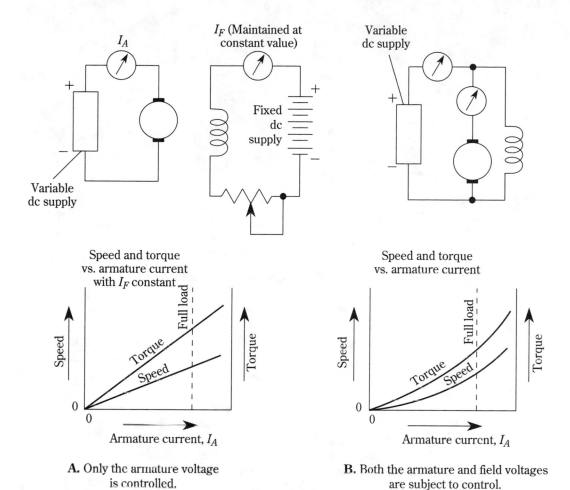

A. Only the armature voltage is controlled.

B. Both the armature and field voltages are subject to control.

2-13 Speed control by nonrheostat variation of armature voltage.

In the control circuits of Fig. 2-13, the fully rated mechanical load is already coupled to the motor shaft. Then, the armature voltage is brought up from zero, with the speed and torque parameters being recorded with respect to increments of armature current.

The series motor

In the series motor, field excitation is derived from windings comprising relatively few turns of heavy wire connected in *series* with the armature. Because of this arrangement, field and armature current are the same. Obviously, the armature and field must experience different interactions in the series motor than in the shunt motor. This is immediately evident in the speed and torque curves of Fig. 2-14. Because the armature and field currents are identical, the basic motor torque equation, $T = k\,\Phi\,I_a$ can be expressed in the form, $T = kI_a^2$ for the series motor. Thus, torque is an exponential function of armature current as shown in the plot for torque versus armature current.

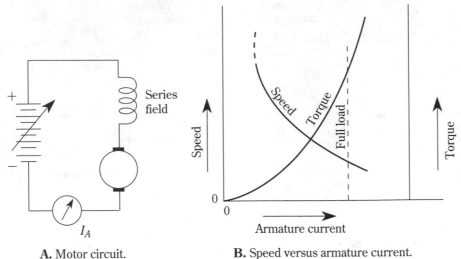

A. Motor circuit.   **B.** Speed versus armature current.

2-14   Basic characteristics of the series motor.

You might already be aware that the series motor is usually selected for applications where high starting torque is needed, such as in traction vehicles. Paradoxically, the torque in the vicinity of standstill and at low speeds appears to be inferior to that of the shunt motor. However, at standstill, a motor is in an *overloaded* condition—there is no counter EMF to impede the inrush of current. It is here that the series motor is advantageous. The initially high armature (and field) current generates a very high torque. Thus, in Fig. 2-14B it is actually the torque depicted to the *right* of the full-load line that accounts for the excellent starting ability of the series motor. (In practice, magnetic saturation sets in and the starting torque is not as high as it would otherwise be.)

The speed characteristics of the series motor also favor those applications where high torque automatically becomes the trade-off for speed. Such a situation is found in traction vehicles and also in many power tools. A shortcoming of the series motor is its *speed-runaway* feature at light loads, and particularly at no load. Small motors can be protected by their own bearing friction and windage, but most series motors tend toward self-destruction if the load is decoupled from the shaft. Unlike the runaway in shunt motors, the accelerated racing of the series motor is accompanied by decreasing armature current. Fuses or circuit breakers in the line, therefore, cannot provide protection against such a catastrophe.

Reversing the polarity of the dc power source does not reverse the rotation of the series motor because both armature and field flux are thereby changed and the magnetic torque remains in the same direction. Reversal must be made by transposing the connections of *either* the armature or the field. By the same reasoning, many series motors will operate fairly well on ac. Optimized ac performance is attained in specially designed series motors, known as *universal motors*.

The compound motor

The compound motor, as its name clearly implies, is a combination type. It incorporates the techniques of both the shunt motor and the series dc motor. Inasmuch as it is both a shunt and a series motor, it might reasonably be expected to display some of the characteristics of each type. This, in essence, is true and accounts for the popularity of compound motors. By appropriate "blending," the no-field runaway behavior of the shunt motor and the no-load runaway characteristic of the series motor can be eliminated. The connections used in compounding are shown in Fig. 2-15. Actually, the difference in performance between the "short" and "long" compounding is generally not of appreciable consequence. Other factors, such as convenience of terminations, reversing considerations, and internal connections to interpole and compensating windings, usually dictate the choice of these near-equivalent connections.

On the other hand, the way in which the shunt and series windings are polarized with respect to one another has a great effect on the operating characteristics. Clearly, the two fields can either aid or oppose in the production of magnetic flux. If the shunt and series fields are connected so that their fluxes *aid*, the motor is said to be *cumulatively compounded*. If their fluxes *oppose*, the motor is *differentially compounded*. These connection techniques have nothing to do with short- and long-shunt connections. In a given machine, if either the series- or the shunt-field connection is transposed, the nature of the compounding changes—if cumulative compounding had previously been used, the transposition converts the machine to the differential type, and vice-versa. A given machine may, or may not, be intended for both types of compounding. Usually, a motor behaves optimally for only one kind of compounding.

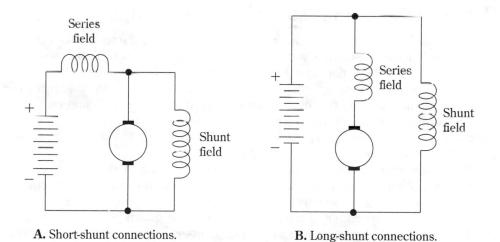

A. Short-shunt connections.   **B.** Long-shunt connections.

2-15   Compound motor connections.

Figure 2-16 shows the torque and speed behavior of compound motors as a function of armature current. These characteristic curves are often considered in terms of their deviation from similar plots for the shunt motor. Because the compound motor merges the features of both shunt and series motors, the curves of Fig. 2-16 depict the behavior of all four motor types.

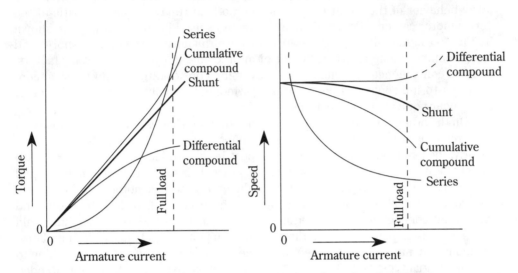

A. Torque as a function of armature current.   **B.** Speed as a function of armature current.

2-16   Torque and speed characteristics of the four basic dc motors.

The user's comparison of motor ratings

In the classification of motor characteristics, one of the first parameters that comes to the mind of the user is the *horsepower* output rating. Although the engineer might have derived his or her basic design from the behavior of a motor with respect to armature current, the buyer will probably be motivated more by the cost relative to the mechanical output. This has already been mentioned in connection with the available torque from a machine. The nameplate on a motor usually specifies its shaft output in terms of *horsepower*, if for no other reason than that most work demand is specified in this way.

The other shaft quantity, *torque*, must be accompanied by a *speed* designation to be meaningful in practical motor situations, So, if both torque and speed are specified, one can also resort to a horsepower rating, which is proportional to the product of torque and speed. Of course, that all-important quantity, starting torque, involves zero speed—here torque is a useful concept even when horsepower output is zero. Therefore, many applications are best met by considering motor behavior as a function of torque.

The torque and the speed characteristics of the four basic dc motors are shown as functions of *shaft horsepower* in Fig. 2-17. These curves are particularly useful for comparing the performance of motors because the plotting assumes that all motors have the same full-load horsepower rating.

A useful feature of the compounding technique is that motor characteristics can be tailored to fit requirements. For example, the speed regulation of a shunt motor can be improved by introducing a small amount of differential compounding—often a very nearly constant speed can be had in this way. Some compounded motors are essentially series-type machines with a small amount of cumulative compounding from a shunt field winding. Other combinations and proportions can be used to fit the purpose at hand.

A word of caution is in order with regard to the differentially compounded motor. This combination is susceptible to racing at high loads because the opposing series field tends to grossly weaken, or even overcome, the field flux produced by the shunt winding. Protective circuitry is indicated here, with armature current as the "sensed" quantity.

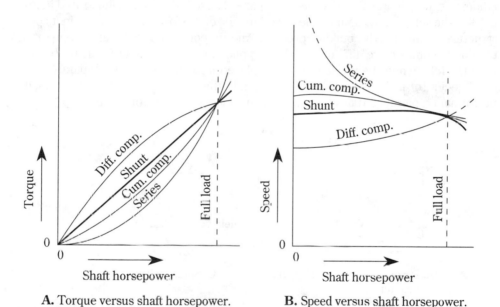

A. Torque versus shaft horsepower. **B.** Speed versus shaft horsepower.

2-17 A user-oriented comparison of the four basic dc motors.

A motor is also a generator

As already pointed out, dc motors and generators are, in principle, very nearly the same. What differences there are have to do with optimization of certain operating features, primarily commutation. Many actual machines are capable of rendering satisfactory service when employed in either function. And, no matter what the "normal" function of a dc machine is, the alternate function is always taking place simultaneously. Thus, you have counter EMF in motors, and you have countertorque in generators. This fact is brought out even more emphatically when dynamic braking is used—that is, when the coasting interval of a motor is shortened by dumping its generated power into a resistor or back into the power line. It follows that a knowledge of generator characteristics is bound to sharpen one's insight into dc machines even if one's primary interest is motors.

The basic dc generators are depicted in Fig. 2-18. The simplest machine, the permanent magnet type shown in Fig. 2-18A, is often the most precise. This accounts for its popular use as a tachometer. In such service, extreme linearity, high-grade construction, and other instrument-like qualities are often incorporated in its design.

The series generator is shown in Fig. 2-18B. This generator cannot build up without a load. Its characteristics are somewhat wild, but nonetheless they are very useful for certain applications. The series field comprises a relatively few turns of heavy conductor, capable of carrying the full armature current. (Resemblance to the series motor is intentional.)

The shunt generator can be operated in two distinct ways—as a self-excited machine, as shown in Fig. 2-18C, or in the separately excited mode, as illustrated in Fig. 2-18D. The self-excited shunt generator depends on *residual magnetism* in its pole structure to initiate the buildup process. And, in contrast to the series generator, buildup becomes more difficult with load application. It is often desirable to *disconnect* the load while the machine is being placed into operation. The separately excited generator involves no buildup problem. In this respect, as well as in others, its performance parallels that of the permanent-magnet generator. ("Build-up" desig-

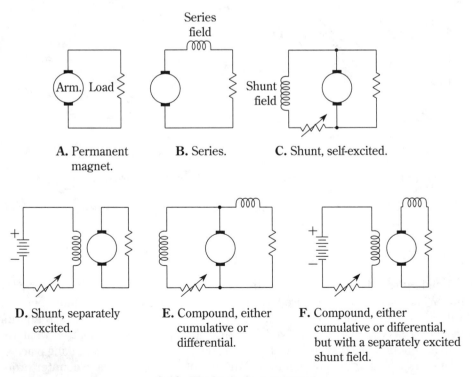

A. Permanent magnet.

B. Series.

C. Shunt, self-excited.

D. Shunt, separately excited.

E. Compound, either cumulative or differential.

F. Compound, either cumulative or differential, but with a separately excited shunt field.

2-18 The basic dc generators.

nates the regenerative sequence of events that enables a generator with no external field excitation to develop full operating capability after being pressed into service. Because of residual magnetism in the field poles, a little bit of armature voltage is developed. This *reinforces* field excitation, which in turn results in *more* armature voltage, etc. After a few seconds, or a few minutes in large generators, the process stabilizes because of magnetic saturation.)

Characteristics of the basic dc generator

The characteristics of the basic generators are illustrated in Fig. 2-19. It is assumed that all machines are driven at a constant speed. As previously mentioned, the plot of terminal voltage versus load current for the series generator, as shown in Fig. 2-19A, is "wild." However, the fact that the terminal voltage increases with load current throughout most of the operating range can be useful. Such a characteristic can provide automatic compensation for the natural voltage drop occurring in long transmission lines. However, a better implementation of this technique results when the series generator characteristic is merged with the shunt generator characteristic. The right-hand portion of the characteristic curve of Fig. 2-19A approximates that of a constant-current source. In this operating mode, the series generator has proved useful as a power source for welding and for certain arc-lamp systems.

The characteristics of compound generators are shown with reference to those of the shunt generator in Fig. 2-19B. The so-called flat compound machine derives its name from the fact that the no-load and rated-load terminal voltages are the same. At other loads, there is a departure from the ideally flat characteristic. Flat-, under-, and over-compounding differ in the amount of series characteristic introduced. The differential-compound characteristic resembles that of the high-load-current region in the series generator, and the applications are similar.

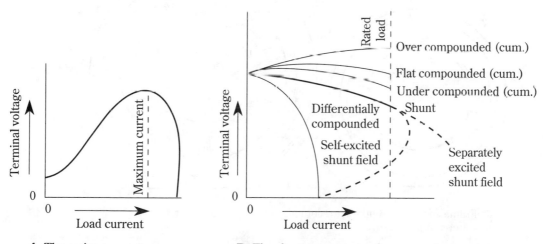

A. The series generator.   **B.** The shunt generator and compound generator.

2-19   Characteristics of the basic dc generators.

The slower a generator is driven, the more field excitation it must have in order to maintain its full-load capability. And, because slower speeds involve operating more deeply into the magnetic-saturation region of the pole structures, the voltage regulation of the shunt and the cumulative compound generators improves. (That of the differential compound generator, already bad, becomes worse; however, its constant-current characteristic *improves*.) Driving a generator below its rated speed might not always prove to be a desirable trade-off because the *efficiency* decreases.

Some unique aspects of modern permanent-magnet motors

At first glance, the characteristics of the shunt-wound motor and the permanent-magnet motor might appear to be similar. It would seem to make little difference to the armature current whether it interacted with a field produced by an electromagnet or a permanent magnet. Indeed, many motor control circuits allow the option of using *either* of these dc motors.

However, there are important differences. Some of these differences are not indicated in older texts and handbooks. For example, it was once commonly held that the permanent-magnet motor was suitable for applications where greater size and weight were not objectionable. Because of modem magnetic materials, this situation is now *reversed*. Figure 2-20 shows the relative frame sizes of a ¼-hp permanent-magnet motor with its ferrite-ceramic poles and a shunt-wound motor of the same horsepower rating. Various alloys of Alnico magnet material are also used in motors.

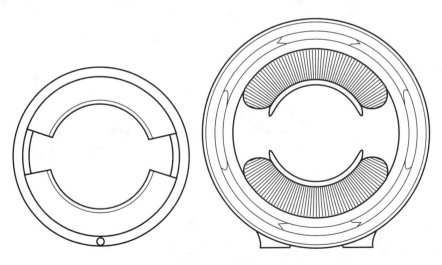

2-20 Frame sizes for a ¼ hp ferrite ceramic permanent-magnet motor and a ¼ hp shunt-wound motor.

In general, the permanent-magnet motor tends to be smaller in size, lighter in weight, more efficient, and more reliable than its shunt-wound counterpart. This statement could not have been made just a few years ago when the first ceramic-field motors were used in toys and other noncritical applications. The earlier steel-magnet motors suffered in reliability because of their susceptibility to *demagnetization*.

Figure 2-21A compares the speed/torque relationships for the two types of motors. The fact that the rated ¼-hp output is developed at somewhat different shaft speeds is not of appreciable consequence in this comparison. Those accustomed to the general nature of the speed/torque relationship in shunt-wound motors are often startled by the greatly extended characteristics of the permanent-magnet motor. For example, in Fig. 2-21A the linear slope of speed regulation continues right down to *standstill* as more torque is extracted by the mechanical load. This implies that the

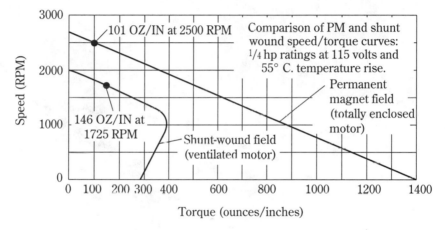

A. Comparison of the permanent-magnet motor and a shunt wound motor.

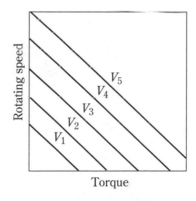

B. Speed/torque relationships of a permanent-magnet motor for various armature voltages.

2-21 Unique characteristics of the modern permanent-magnet motor.

permanent-magnet motor has a *starting-torque* capability several times that of its shunt-wound counterpart. Additionally, the speed as a function of load is easier to predict. This is even more clearly shown in Fig. 2-21B where the armature is subjected to various voltages.

The surprising performance differences in permanent-magnet motors stem from the much smaller effect of armature reaction on field strength. The high coercive force of modem magnetic materials is primarily responsible for this.

3
The classic ac motors

HAVING REVIEWED THE CHARACTERISTICS OF THE CLASSICAL DC MACHINES IN THE LAST chapter, this chapter will give the same treatment to the classical ac motors. The discussion begins with the ac series motor. As previously mentioned, the ordinary dc series motor can often give a satisfactory account of itself when operated from the ac power line. However, much better efficiency and commutation are obtained when a series motor operated from an ac source is designed for ac use. It then performs well on dc also. Such a machine is known as a *universal* motor. Because of economic factors, not all universal motors make use of all the design and structural techniques that are known to optimize ac operation. This is particularly true for small fractional-horsepower sizes and reflects the general trend for series motors under ¼ horsepower.

Operation of the series motor from ac power

Compared with dc series motors, the ac type can differ in at least some of the following respects:

- The ac motor tends to have fewer series-field ampere turns but tends to have a higher number of ampere turns on the armature.
- The field structure of the ac motor is highly laminated to reduce eddy current loss.
- In larger sizes (greater than ¼ horsepower), the ac-type series motor uses distributed, rather than concentrated, poles.
- In the larger machines, the ac-type series motor is likely to have a greater number of poles.
- When compensating windings are used, they are often *inductively*, rather than conductively, associated with the armature. That is, these windings are

short circuited and are not physically connected in the motor circuit.

- In the larger sizes, resistance is sometimes introduced between the armature conductors and the commutator segments. This is a commutation aid, and protects the brushes from high short-circuit currents during starting.
- In the ac series motor, there is more emphasis on making the air gap small.

Most series motors of less than ¼-horsepower rating are classified as universal motors if their design permits satisfactory operation on ac up to 60 Hz. Motors with larger ratings are often referred to simply as ac series motors. Unless specially designed, most small fractional-horsepower universal motors will not develop the same speed and torque characteristics on both ac and dc. For a large number of applications, however, the behavior will be sufficiently similar to satisfy requirements. If operation is desired at frequencies above 60 Hz, custom designing is usually involved.

Some practical aspects of ac series motors

Figure 3-1 illustrates some of the practical aspects of series motors. The schematic diagram shown in Fig. 3-1A is the same as that for a dc series machine, except for the connection to an ac source. Notice that a double-pole, double-throw switch would be needed to accomplish reversal of rotation (either the field terminals or the armature terminals would have to be transposed). The connection of compensating and interpole windings is shown in Fig. 3-1B. Here, again, the technique duplicates that used in dc series motors.

The ac series motor depicted in Fig. 3-1C also has compensating and interpole windings. However, these auxiliary windings, rather than being conductively connected to the armature circuit, are *inductively coupled*. In ac applications, such a technique produces beneficial results. First, the armature can develop more torque because its applied terminal voltage is not diminished by the reactive voltage drops that would occur across these auxiliary windings if they were conductively connected to the armature circuit. Second, it is found that the power factor of the ac motor is improved.

The general concept of *transformer action* inside motors is relevant to control techniques. In the dc motor, the ac-carrying armature does not induce currents in the main field windings. This is because the magnetic flux from the main field poles is in *space quadrature* with the armature flux. These two magnetic fields interact to be sure, but the magnetic lines of force from the armature do not link the turns of wire in the main field poles. The situation is analogous to that prevailing when an ac-carrying solenoid is oriented at right angles to the longitudinal axis of a second solenoid. Despite the physical proximity of the two solenoids, the second one will not become a "secondary" current source by transformer action.

To facilitate reversal of rotation, some series motors are equipped with two main field windings. These are polarized, or phased, to produce opposite rotation—only one such winding is active when reversal is accomplished with a single-pole, double-throw switch as shown in Fig. 3-1D. The characteristic curves of a typical small universal motor are shown in Fig. 3-2. At 60 Hz, ac series motors generally exhibit the highest horsepower-to-weight ratio of any ac motor. This is largely because of the high speed at which these series motors can operate.

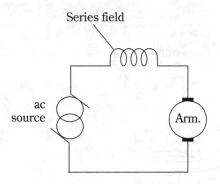

A. General circuit connection for ac series motor, or universal motor.

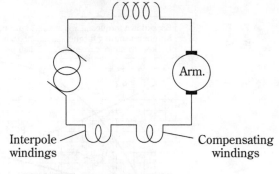

B. Ac series motor, or universal motor, with conductively connected compensating and interpole windings.

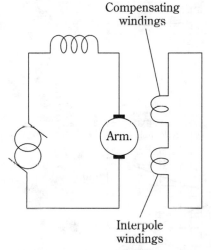

C. ac series motor, or universal motor, with inductively coupled compensating and interpole windings.

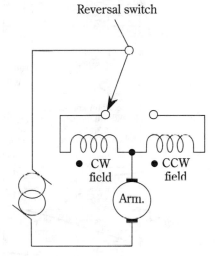

D. Reversible ac series motor, or universal motor, with split field windings.

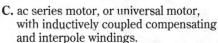

3-1 The ac series motor.

The repulsion motor

The discussion of the ac series motor, or universal motor, will serve to introduce the basic principle underlying the *repulsion motor*. The use of resistance inserted between the armature conductors and commutator segments was mentioned. Such resistance limits the short-circuit current that flows between adjacent commutator segments that happen to be located under the brushes during a start from standstill.

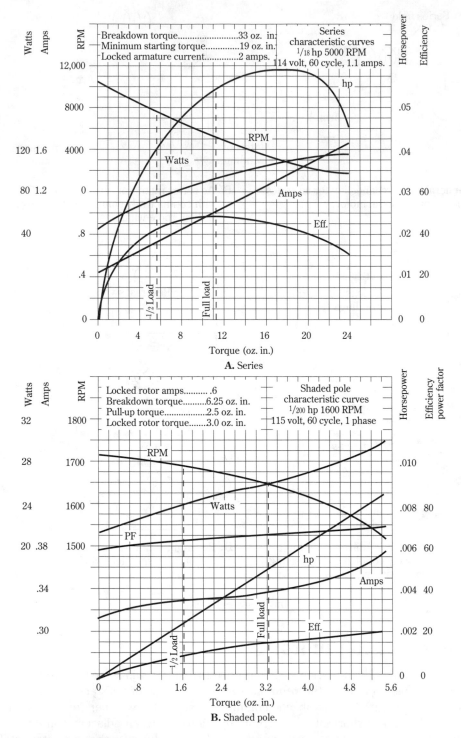

Watts Amps RPM

Breakdown torque.........................33 oz. in.
Minimum starting torque..............19 oz. in.
Locked armature current.............2 amps.

Series
characteristic curves
$^1/_{18}$ hp 5000 RPM
114 volt, 60 cycle, 1.1 amps.

Horsepower Efficiency

12,000

hp

8000

.05

RPM

120 1.6 4000

.04

Watts

80 1.2 0

.03 60

Amps

40

.8

Eff.

.02 40

.4

.01 20

$^1/_2$ Load

Full load

0

0 0

0   4   8   12   16   20   24

Torque (oz. in.)

A. Series

Watts Amps RPM

Locked rotor amps............6
Breakdown torque..........6.25 oz. in.
Pull-up torque.................2.5 oz. in.
Locked rotor torque.......3.0 oz. in.

Shaded pole
characteristic curves
$^1/_{200}$ hp 1600 RPM
115 volt, 60 cycle, 1 phase

Horsepower Efficiency
power factor

32 1800

28 1700

RPM

.010

24 1600

Watts

.008 80

20 .38 1500

PF

.006 60

hp

.34

Amps

.004 40

.30

Full load

Eff.

.002 20

$^1/_2$ Load

0   .8   1.6   2.4   3.2   4.0   4.8   5.6

0   0

Torque (oz. in.)

B. Shaded pole.

3-2 The ac characteristics of a small universal motor. Bodine Electric Co.

The source of this short circuit current is unintentional transformer action between the field windings and the armature conductors (this did not exist in the stationary dc series motor). It contributes nothing to motor action but is very hard on the brushes and the commutator. In the running motor, the same action continues to take place, but its effects become less harmful because the heating is then distributed over all of the commutator segments. This undesirable effect stems from *transformer action*.

A clearer case of transformer action accounted for the effect of inductively coupled compensating and interpole windings. Here, the effect was achieved by intent.

It would be only natural now to ponder the feasibility of coupling ac power into the rotating armature via *electromagnetic induction* rather than via the conductive connection to the field windings used in series motors. True enough, it has just been stated that this already occurs, causing heat dissipation. However, it is desirable to make the induced current somehow develop *torque*, rather than primarily heat. You can "invent" this motor by the simple expedient of short-circuiting the two brushes and rotating the brush axis so that it coincides *neither* with the geometric-neutral axis nor with the axis of the poles. And the conductive connection between series field and armature is eliminated.

The repulsion motor thereby created retains the basic speed and torque behavior of the ac series, or universal, motor. It has two important advantages, however. It can be designed so that the objectionable dissipative transformer current is virtually neutralized at a certain speed, thereby providing relatively clean commutation. In practice, good commutation is obtained over a usable speed range. And, because the armature is *not* connected to the power line, it can be designed for a convenient low voltage. This greatly reduces insulation problems in the armature and in the brush assembly. At higher speeds, the dissipative transformer currents provide a braking effect, which prevents racing at low speeds. Figure 3-3 shows the basic relationships pertaining to the repulsion motor.

If it has not previously been encountered, the shorted armature depicted in these diagrams might be startling. The symbolization is somewhat modified from that used before, in order to emphasize that the field poles are alternately north and south—the phase dots reveal that the field windings are connected so that when one pole is north, the other pole is south. And, unlike the armature in ac series and universal motors, the armature in repulsion motors is conductively isolated from both the field and the power line.

In the situation shown in Fig. 3-3A, the short-circuit brushes are analogous to a short across the detector of a balanced Wheatstone bridge. That is, the voltages induced in the two sectors of the armature winding are equal and are polarized so that there is no net current flow through the shorting connection. This is tantamount to saying that no current flows in the armature conductors. Consequently, no torque is developed.

In Fig. 3-3B, the brush axis coincides with the axis of the field poles. If the necessary modifications to accomplish this were made, considerable humming, vibration, and heat would be produced. However, no net torque would be forthcoming because now the two sectors of the armature winding carry equal, but opposite, currents. The resulting torques, though strong, cause rotation with equal force in opposite directions.

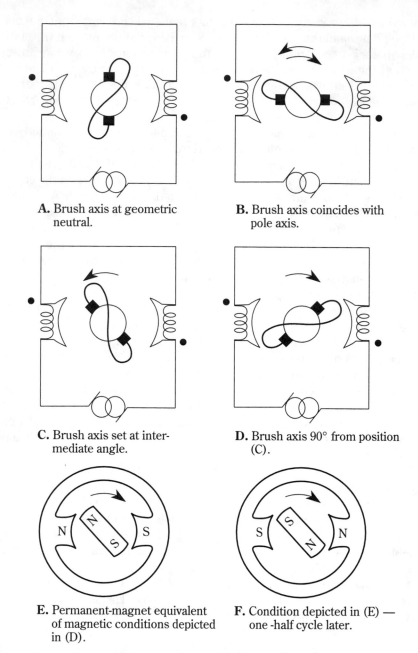

A. Brush axis at geometric neutral.

B. Brush axis coincides with pole axis.

C. Brush axis set at intermediate angle.

D. Brush axis 90° from position (C).

E. Permanent-magnet equivalent of magnetic conditions depicted in (D).

F. Condition depicted in (E) — one-half cycle later.

3-3 Basic relationships in the repulsion motor.

The in-between brush positions shown in Figs. 3-3C and D enable motor operation to be achieved because of the unequal currents that are developed in the two sectors of the shorted armature. Torque in one direction then predominates, and the motor operates and carries a load. The application of a mechanical load slows down the speed, enabling the current difference between the armature segments to increase because of a decrease in generator action. Accordingly, a greater electromagnetic torque is developed (the interaction between transformer and generator action is similar to that between counter EMF and applied armature voltage in a dc machine). Notice that the direction of rotation is the same as the direction of brush shift from geometric neutral.

In Figs. 3-3E and F, the permanent-magnet equivalents of the situation in Fig. 3-3D are shown for both halves of the ac cycle. In the rotating motor, successive armature loops occupy an angular position such that their magnetic field simulates that of the permanent-magnet rotor. Whereas the magnet rotor can experience only a transitory displacement from its illustrated position, the motor armature rotates continuously because a new "magnet" is introduced every time appropriate armature loops undergo shorting action by the brushes. The speed-torque characteristics of the repulsion motor are shown in Fig. 3-4.

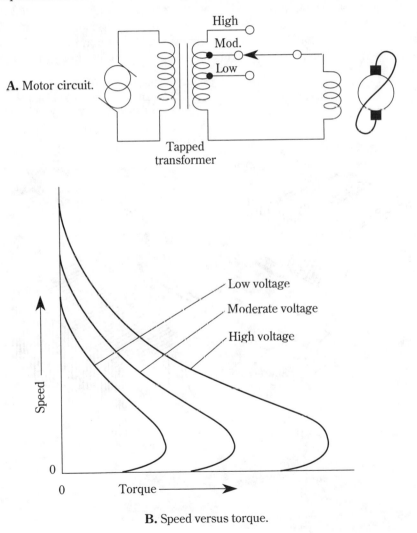

A. Motor circuit.

High
Mod.
Low

Tapped
transformer

Low voltage

Moderate voltage

High voltage

Speed

0

0   Torque

B. Speed versus torque.

3-4 General speed/torque characteristics of the repulsion motor.

The single-phase induction motor

In the design and construction of large *transformers*, consideration must be given to the mechanical force existing between the primary and secondary windings. This is particularly important in order to ensure reliability in the event of a short circuit on the secondary winding. Otherwise, even before thermal damage could occur, the transformer might be catastrophically destroyed by the physical force tending to move the two windings farther apart. It will be recognized that this is a manifestation of Lenz's law—the induced magnetic field opposes the inducing field. Being mindful of this, we might try to convert the tendency towards lateral motion in the windings of short-circuited transformers into continuous *rotational* motion. This would appear to be a simple enough task, and we might produce the device shown in Fig. 3-5.

In Fig. 3-5, the "primary" of our "rotary transformer" is the stator of the machine, whereas the short-circuited "secondary" is the *rotor*. This looks promising because no commutator or brushes are needed. However, upon energizing the stator from an ac source, you might have become disappointed with the humming and heating of this strange transformer and might conclude that rotation was not forthcoming.

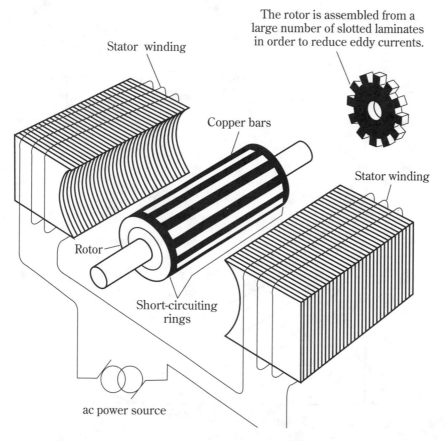

3-5 The squirrel-cage induction motor can be viewed as a "rotary transformer."

To start rotation of the induction motor, manually spin the shaft. The motor then will accelerate up to operating speed, and even a relatively crude experimental model will reveal the potentially favorable features inherent in an ac motor of this type.

If the single-phase induction motor was not set in operation, an accurate report of the experiment would have stated merely that no net torque was developed in the rotor. The fact that equal, but opposite, torques were produced would have been suggested by the vibration and humming.

A mathematical approach is helpful in understanding the torque in the single-phase induction motor. Figure 3-6A is a simple vectorial representation of the field "seen" by the rotor when the starter is fed with a sinusoidally varying current. The vertical projection of the rotating vector depicts the magnetic field strength, Φ, as an angular function of maximum field strength, Φ_{max}. The cyclic excursion of Φ_{max} is made in response to the sinusoidally varying current in the stator. If, for example, the vector diagram represents a time corresponding to 45° in the progress of such a sine wave, then Φ will have an amplitude equal to sine 45° (Φ_{max}), or 0.707 Φ_{max}. However, there is more than meets the eye here.

In Fig. 3-6B, an equivalent vectorial representation of the sinusoidally varying field is shown. In this diagram, field strength Φ is produced as the result of counter-rotating fields, each having the value of

$$\frac{\Phi_{max}}{2}$$

Again, Φ is always a vertical vector. Here Φ goes through the same amplitude variations as it does in Fig. 3-6A. This being so, it must be an equivalent graphical and mathematical way of showing how the pulsating field, Φ, varies in amplitude during the excursion of a cycle of stator current.

The significance of the counterrotating vectors,

$$\frac{\Phi_{max}}{2}$$

is shown by the curves in Fig. 3-6C. These curves reveal that two equal and opposite torques prevail during standstill. Obviously, there is no reason for the motor to rotate in one direction or the other. However, when a small rotational force is imparted to the rotor from an outside source, one torque exceeds the other. It can be assumed that the shaft is given a spin by some manual method, such as pulling a cord wound several times around a pulley. When a net torque exists in one direction, the motor will accelerate until it approaches synchronous speed. Synchronous speed is given by the equation:

$$S = \frac{120 \times f}{p}$$

where:

S is the synchronous speed in revolutions per minute,
f is the frequency of the ac power source in hertz, and
p is the number of poles in the stator.

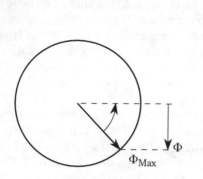

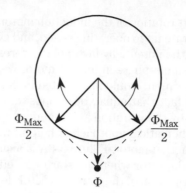

A. Simple vector representation of a sinusoidally varying field.

B. An equivalent representation in which two counterrotating vectors produce the resultant field.

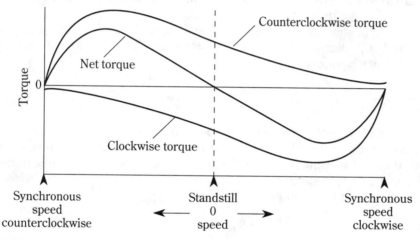

C. Rotor torque as a function of speed.

3-6 Field strength and torque in the single-phase motor.

   The motor can never attain synchronous speed because, at that speed, the net torque is zero. The difference between synchronous speed and actual speed is called the *slip* speed. The more heavily loaded the motor is, the greater the slip speed is. This behavior constitutes reasonable proof for the theory of counterrotating fields and their accompanying torques.

   Another way of describing the action of the rotor is to say that once it is rotating, it is under the influence of a rotating field that is stronger than that of the oppositely rotating field. The concept of a rotating field is important because all induction motors produce a rotating field. In our discussion, an external starting method was used to expose the rotor conductors to a net rotating field. Naturally it is desirable to have a motor that is able to start on its own. There are various starting arrangements and devices that pertain to many differently named motors, all of which are essentially induction motors.

The split-phase induction motor

Besides the main field winding on its stator, this squirrel-cage induction machine also has an auxiliary, or starting, winding. The starting winding is physically displaced at an angle corresponding to 90 electrical degrees from the main field winding. Additionally, the starting winding comprises *fewer* turns of *smaller-gauge* wire than the main field winding. The starting winding will have a low reactance but a *high resistance* relative to the main field winding. The objective is to produce a second field that is both in space and in time quadrature with respect to the main field. Such a composite field simulates that produced by a two-phase power source and rotates at a synchronous speed. The rotation of the rotor occurs only when, by one means or another, it experiences the effect of a rotating magnetic field.

Part of this objective is readily attainable. The space-quadrature requirement is realizable from the physical orientation of the starting winding relative to the main field winding. The 90° phase displacement between the currents in the windings is not attained because the starting winding is not a perfect resistance and the main field winding is not a perfect reactance. Nonetheless, the composite field has a *component* that rotates, and this suffices to develop starting torque in the rotor. Somewhere in the vicinity of about 80 percent of synchronous speed, a shaft-mounted centrifugal switch opens the starting-winding circuit. If the starting-winding were to remain connected, its torque contribution would be minimal, but its energy dissipation could lead to an unsafe rise in temperature. The equivalent circuit and the phase relationships in this motor are shown in Fig. 3-7.

The operating characteristics of a small split-phase induction motor are shown in Fig. 3-8. Notice that the speed behavior is similar to that of a dc shunt motor. As the torque demand is increased, the motor slows down and is able not only to accept higher current but to operate at a higher power factor from the ac line. Both of these changes lead to the development of greater torque in the rotor. The current induced in the rotor is largely due to its relative motion with respect to the rotating field. This relative motion is a function of the slip, which is the difference between the actual speed of the rotor and the synchronous speed. Although the mechanism is somewhat different, both the dc shunt motor and the induction motor slow down in order to accept more torque-producing current when the mechanical load is increased.

Control features
of the split-phase induction motor

Limited speed control can be achieved in the split-phase induction motor. Control of the slip is accomplished by varying the magnitude of the voltage applied to the stator. However, the speed must not be made so low that the centrifugal switch allows the starting winding to be energized during normal operation. Controlling the speed

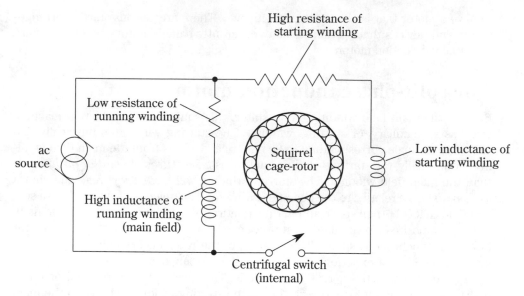

A. Equivalent circuit.

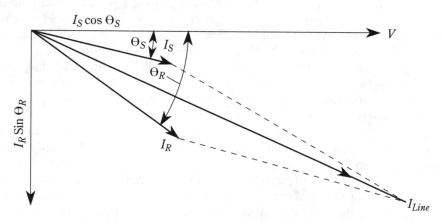

B. Initial phase relationship during starting interval. I_S is the current in the starting winding, and I_R is the current in the running winding.

3-7 The resistance-start, split-phase motor.

by varying the *frequency* of the power supply enables higher speeds to be reached than if the rotor slip were controlled. However, the same low-speed limitation exists because of the centrifugal switch. Clever implementation of electronic control and logic circuits can circumvent limitations such as those described above. For example, a tachometer or stroboscopic speed sensor could actuate logic circuitry so that the starting winding would not be energized when the motor was slowed down from

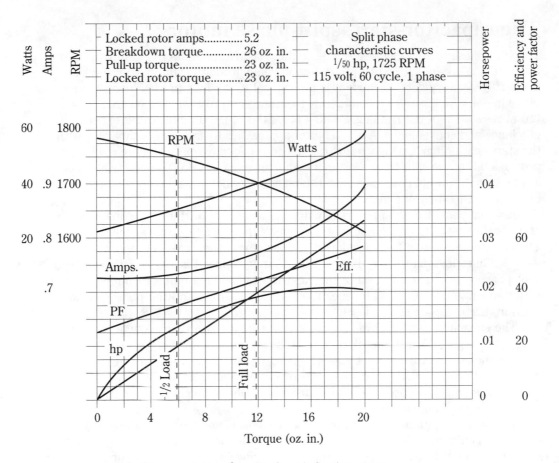

Watts | Amps | RPM

Locked rotor amps.............5.2
Breakdown torque............26 oz. in.
Pull-up torque.....................23 oz. in.
Locked rotor torque...........23 oz. in.

Split phase
characteristic curves
1/50 hp, 1725 RPM
115 volt, 60 cycle, 1 phase

Horsepower | Efficiency and power factor

Torque (oz. in.)

3-8 Characteristics of a split-phase induction motor. Bodine Electric Co.

a higher operating speed (you would have to bring out extra leads from the centrifugal switch in order to make such a modification in a conventional motor).

In general, loads that are primarily inertial in nature tend to be better suited to slip control of speed than loads that are essentially frictional. Fans, blowers, and flywheels exemplify loads with high inertial components. The frequency-control method is often an excellent way to control the speed of induction motors, regardless of the type of load. However, as the frequency is changed, it is generally necessary to also change the applied voltage in direct proportion to the frequency. The split-phase motor can be reversed by transposing the two leads of either the starting winding or the main field (running) winding.

There are split-phase motors in which the auxiliary winding is designed so that it is *always* connected in parallel with the main field winding. There is thus no need for the centrifugal switch. However, it is not easy to incorporate a very high starting torque when this simplification is used. To distinguish between these two types of split-phase induction motors, the one with the centrifugal switch is referred to as a *resistance-start* type.

The capacitor-start, split-phase motor

The split-phase motors just described use the resistive and inductive components of two windings to simulate a two-phase rotating field inside the stator. The simulation is actually not very good because it is impractical to produce predominantly inductance in one winding and predominantly resistance in the other. The insertion of external resistance in the running winding helps this situation, but at the expense of efficiency and convenience. A much better approach is to associate a capacitor with the starting winding. It is known that resistance-capacitance circuits more readily approach a 90° phase shift than resistance-inductance circuits do. The practical capacitor is a much "purer" reactance than the practical inductor, whether the latter is an inductive component or a motor winding.

It so happens that an appropriate capacitor associated with the starting winding of a resistance-start, split-phase motor converts the machine to a capacitor motor. The two-phase rotating field developed in the stator of the capacitor motor is much more symmetrical than the rather "lopsided" one produced in the resistance-start, split-phase motor. The addition of the capacitor results in a quieter running motor with a much greater starting torque. The equivalent circuit and the phase relationships in this motor are shown in Fig. 3-9.

The capacitor-start, split-phase motor is much easier to reverse than the resistance-start type is. To accomplish reversal, the motor is temporarily disconnected. As soon as the speed drops so that about 20 percent slip exists, the centrifugal switch will close its contacts. If, after this occurrence, the motor is re-energized with its starting winding transposed, deceleration and subsequent acceleration in the opposite direction will take place.

The salient features of the capacitor-start motor is its inordinately high starting torque as compared with other "artificial-start" induction motors. Starting torques from three to five times the rated torque are typically found in these motors.

The permanent-capacitor split-phase motor

The permanent-capacitor, split-phase motor makes use of two identical stator windings and a capacitor selected for optimum running characteristics. Because no centrifugal switch is employed, the overall reliability of this motor exceeds that of machines dependent on such a switch. Because of electrical and magnetic symmetry, this motor provides quieter operation than the resistance-start or capacitor-start types. The rotor of the permanent-capacitor motor experiences a rotating field that is similar to the field that results from an actual two-phase power source. In contrast, the resistance-start and capacitor-start motors develop a field with a large pulsating component—this contributes to noise and vibration, but not to torque.

Figure 3-10 shows the circuit diagram for the permanent-capacitor motor. The equivalent series resistance of the winding is not shown because it is not an important

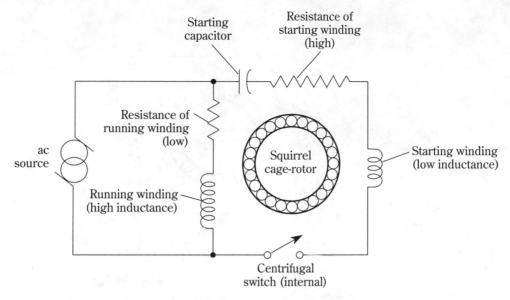

A. Equivalent circuit.

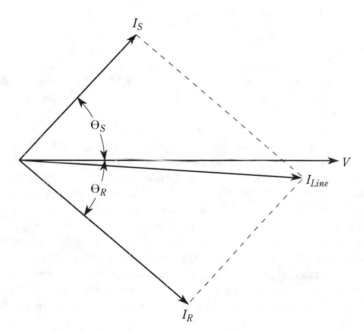

B. Initial phase relationships during starting interval.

3-9 The capacitor-start, split-phase motor.

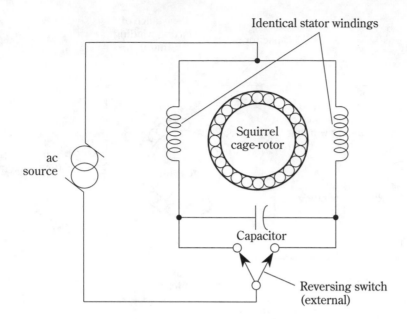

3-10 The permanent-capacitor, split-phase motor.

factor in the basic principle of this motor. Rather, the windings are alike both in inductance and resistance. Both windings are, therefore, "running" windings. Reversal of rotation is very easily accomplished by the use of the single-pole, double-throw switch, as shown in Fig. 3-10. The characteristics for a typical motor of this type are shown in Fig. 3-11.

An outstanding characteristic of this motor is its speed controllability. This is readily achieved by varying the applied voltage. When the applied voltage is greater than the rated voltage, the permanent-capacitor motor, like other induction machines, can neither attain synchronous speed nor exceed it. A high applied voltage only increases the starting torque and improves the speed regulation. However, when the applied voltage is lowered, a wide range of speeds below the synchronous speed is available by virtue of the high slip allowed by this motor.

The permanent-capacitor motor operates at a high power factor, and its overall reliability is enhanced by the fact that the capacitor is usually the oil-filled type, rather than the electrolytic type employed in capacitor-start motors. The trade-off for the many excellent characteristics of this motor is its relatively low starting torque. Most motors of this type have starting torques limited to 50 to 100 percent of their rated torque.

The two-value capacitor motor and the autotransformer capacitor motor

The starting torque of the permanent-capacitor, split-phase motor can be greatly increased by the use of a capacitor whose value is much larger than that suitable for optimum running operation. It is only natural, then, that the two-value capacitor mo-

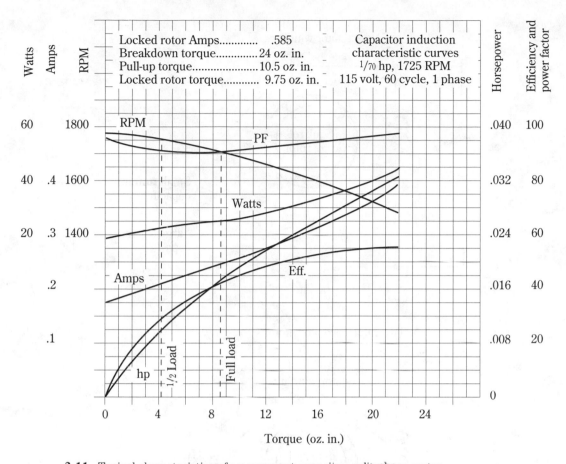

Locked rotor Amps............. .585
Breakdown torque.............24 oz. in.
Pull-up torque.....................10.5 oz. in.
Locked rotor torque........... 9.75 oz. in.

Capacitor induction
characteristic curves
$1/70$ hp, 1725 RPM
115 volt, 60 cycle, 1 phase

3-11 Typical characteristics of a permanent-capacitor, split-phase motor. Bodine Electric Co.

tor is used where high starting torque and quiet operation at a high power factor are desired. The most obvious way to achieve this combination of desirable characteristics is to incorporate a centrifugal switch such as is used in the resistance-start and capacitor-start, split-phase motors. Starting is accomplished with the large electrolytic-type capacitor connected in parallel with the smaller oil-filled running capacitor. In the vicinity of 80 percent of synchronous speed, the centrifugal switch disconnects the starting capacitor. The motor then operates as a permanent-capacitor, split-phase type. The two stator windings are generally identical in these motors. The circuit for the two-value capacitor motor is shown in Fig. 3-12A.

An oil-filled capacitor tends to have a much longer life than an electrolytic type does, when subjected to the rather abusive conditions of motor starting. Therefore, some motors use a single oil-filled capacitor in conjunction with an autotransformer to attain the same effect found in the two-value capacitor motor. During the start interval, the centrifugal switch connects an autotransformer in the circuit so that the actual capacitance is effectively multiplied by the square of the turns ratio of the transformer windings. This can be recognized as an impedance-transforming technique. The capacitor is momentarily subjected to a stepped-up voltage, but these ca-

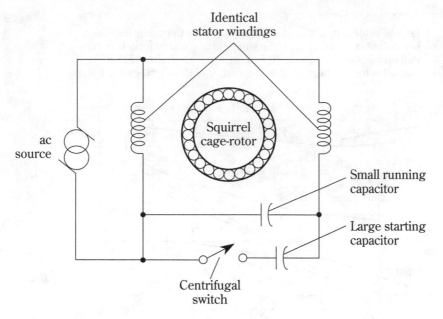

A. Two-value capacitor motor.

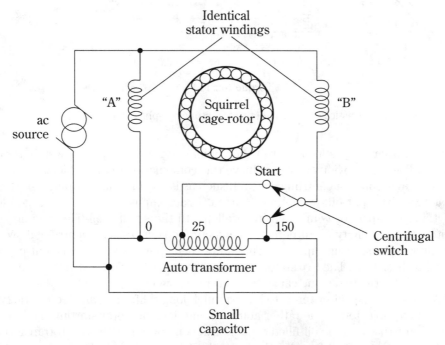

B. Single-value capacitor and auto transformer.

3-12 Two methods for obtaining high starting torque and smooth running.

pacitors are readily designed for high voltage ratings. The overall scheme provides better reliability than that ordinarily obtained with electrolytic capacitors.

If the capacitor in Fig. 3-12B is a 10-microfarad, oil-filled type, and the auto-transformer has 150 total turns, with the tap at 25 turns, the capacitance "seen" by the motor during the starting interval will be

$$\left(\frac{150}{25}\right)^2 \times 10, \text{ or } 360 \text{ microfarads}$$

The capacitor voltage, assuming a 120-volt line, would be

$$\frac{150}{25} \times 120, \text{ or } 720 \text{ volts}$$

In practice, a 1000-volt capacitor would be used.

The shaded-pole motor

The basic construction features of the shaded-pole induction motor are illustrated in Fig. 3-13A. As can be seen, the poles are divided into two segments. The smaller segment of each pole has a shorted turn consisting of a heavy copper strip or bar. The growth and decline of magnetic flux is not uniform across the pole faces, as it is with ordinary pole structures having uniform geometries. Because of the shorted turn, the flux in the small segment always opposes the flux change occurring in the large segment. This is in accordance with Lenz's law—current is induced in the shorted turn, and the field thus developed opposes the inducing field (that is, the field in the large segment). This produces a *delayed* field in the small segment. The net result when the pole structure is excited from a sinusoidal source of current is a *sweeping* action of flux across the faces of the poles as shown in Fig. 3-13B.

Because *both* poles are behaving in this fashion, but in alternate sequence, the rotor of the motor experiences a *rotating* field. Like other motors illustrated in this book, the shaded-pole type can have many pairs of poles, the two-pole structure is shown for the sake of simplicity. The direction of rotation is toward the shaded-pole segment. In most of these motors, the shorted turn is an integral feature of the pole structure and the motor cannot be reversed. However, special models are manufactured in which the poles have "shaded" segments at each pole tip; these, instead of being short-circuited, are electrically switchable.

Although this appears to be a clever way of creating a rotating field from a single-phase field excitation, such motors have relatively weak starting and running torques. Also, their efficiency and power factor are low. Thus, the shaded-pole motor is made only in small sizes and is suitable for only low-torque applications such as phonograph drives, fans, and various instrumental purposes. On the plus side, its simplicity and reliability can hardly be surpassed. There are no internal switches, and the stalled-rotor current is not appreciably greater than its rated full-load current.

The shaded-pole motor can tolerate large amounts of slip before breakdown torque is experienced. It, therefore, is well suited to speed control by variation of applied voltage. As with all induction-type motors, however, synchronous speed can be approached but not attained. The characteristics of a small shaded-pole motor are shown in Fig. 3-14.

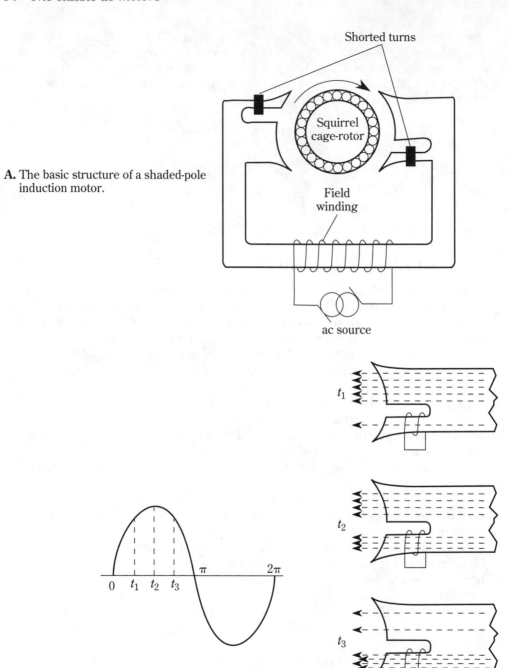

A. The basic structure of a shaded-pole induction motor.

B. The flux-sweeping action of the shaded pole.

3-13 The shaded-pole induction motor.

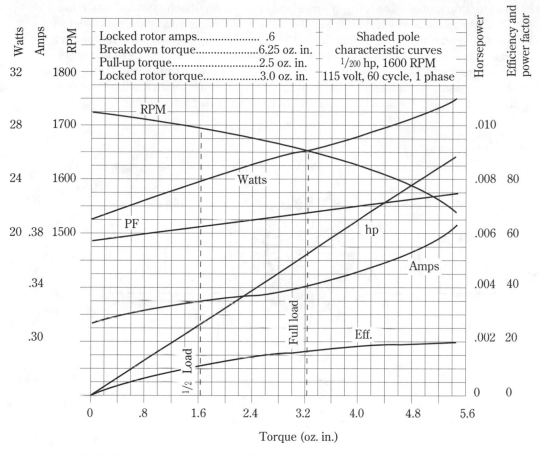

3-14 Characteristics of a small shaded-pole induction motor. Bodine Electric Co.

The repulsion-start induction motor

The important characteristics of the repulsion motor and the induction motor are essentially complementary—the repulsion motor has a very high starting torque but poor speed regulation in the normal operating range, whereas the basic single-phase induction motor develops no starting torque but is a near-constant speed machine once it is operating with normal loads. Therefore, hybrid machines utilizing both motor principles have evolved.

It so happens that, if the commutator segments of a repulsion motor are short-circuited, the previously commutated armature winding simulates quite closely the squirrel-cage rotor of an induction motor. This being the case, a repulsion motor designed with a centrifugally actuated shorting mechanism can start as a repulsion motor and, at about 80 percent of synchronous speed, can shift to the operating mode of an induction motor. This is the appropriately named repulsion-start motor. The basic idea is illustrated in Fig. 3-15. Pivoted copper segments, represented by the radial arrows, are kept parallel to the shaft by a spring when the motor is at standstill or is rotating below 80 percent of synchronous speed. As the motor accelerates, centrifugal force ultimately overcomes the force exerted by the spring and the pivoted segments move outward to make physical contact with the inner surface of the commutator. The machine then operates as an *induction motor*.

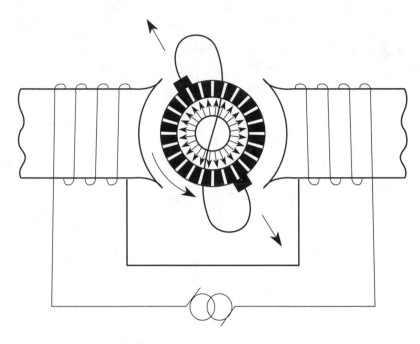

3-15 The repulsion-start induction motor.

The characteristics of the repulsion-start induction motor are shown in Fig. 3-16. Notice that the transition between the two modes of operation takes place at the speed where the induction-motor torque exceeds that of the motor operating in the repulsion mode. Often, the centrifugal mechanism also lifts the brushes from contact with the commutator at the same time that the short-circuiting action occurs. This action extends the life of both the commutator and the brushes.

Reversal of rotation is accomplished by shifting the brush axis. Some designs incorporate separate field windings in order to facilitate reversal. The repulsion-start induction motor is made in both fractional- and integral-horsepower sizes. It is a capable workhorse where polyphase power is not available, but it tends to be noisy, expensive, and it requires relatively high maintenance. An additional negative factor in some applications is the high level of RFI generated during the starting interval. This can play havoc with communications and video equipment, and requires due consideration whenever digital-logic circuits are situated nearby.

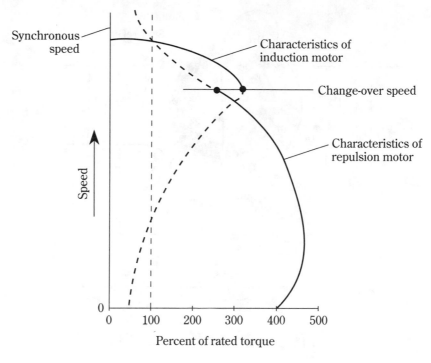

3-16 Characteristics of the repulsion-start induction motor.

The repulsion-induction motor

The repulsion-induction motor can be called a sophisticated version of the repulsion-start motor just described. Approximately the same overall objective is achieved, without the centrifugal mechanism. And, other than the commutator and short-circuited brushes there is no switching process involved as the motor accelerates. Also, the brushes are never lifted from contact with the commutator. The basic design of this simplified machine is shown in Fig. 3-17. The brushes, commutator, and the commutated armature winding are arranged much as in a repulsion motor. Deeply embedded in the armature iron, however, is a squirrel-cage type winding with short-circuiting rings welded at each end. Recalling that the stators of single-phase repulsion and induction motors are basically the same, it is natural to ponder which motor mode predominates in this hybrid machine.

During the starting interval, the slip frequency is initially equal to that of the *power line*. Inasmuch as the squirrel-cage winding has deliberately been made highly inductive, its reactance impedes the flow of short-circuit current. Under these conditions, the squirrel cage contributes very little torque as long as the motor speed remains a small fraction of synchronous speed. So, initially, the rotation of the armature is produced by the torque developed in the commutated armature winding. This is desirable, because the repulsion motor is a star performer in the torque department and is capable of exerting brute turning effort in the vicinity of zero speed.

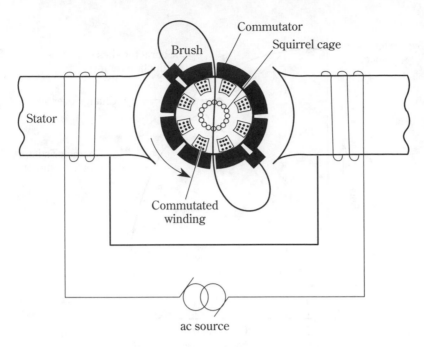

3-17 The repulsion-induction motor.

As the armature accelerates, the *frequency* of the current induced in the squirrel cage decreases. This results in less inductive reactance and, therefore, *greater* torque-producing current. At the same time, the repulsion-motor torque is decreasing, this being the "nature of the beast." Refer to the repulsion-motor speed-torque curve of Fig. 3-4. Somewhere in the vicinity of 80 percent of synchronous speed, induction-motor action begins to predominate. Therefore, the speed-regulation curve departs from what it would be for a repulsion motor and assumes the flatter characteristic generally associated with induction motors. An exception is the speed range above the synchronous speed. But, how can an induction motor perform in this speed range?

It is still true that induction motors can only approach synchronous speed. Certainly, they cannot exceed it. The existence of a supersynchronous speed range in the repulsion-induction motor is due to the fact that the repulsion-motor characteristics continue to exert influence even at zero torque demand. Thus, if the shaft of this motor is spinning freely, without any external load, the torque developed by the commutated-armature winding boosts the speed above its synchronous value. However, the extent of this action is limited because of countertorque developed by the squirrel-cage winding—it functions as an induction generator above synchronous speed. The unique behavior of this machine stems from the fact that the inductive reactance responsible for the interchange of motor characteristics does not display the abrupt and positive action of an electrical switch.

By the same rationale, inductive motor action is present down to zero speed. Indeed, there is always an interchange of energy between the two windings. This *coupling* between the windings makes the starting torque slightly less than that obtained in the switch-type repulsion-start motor previously discussed. However, the power factor and the commutation tend to be improved by the presence of the two windings.

Unlike other induction machines, the repulsion-induction motor has the desirable features of a speed-torque characteristic that is not subject to "breakdown" by temporary overloads. The speed-torque curve is shown in Fig. 3-18.

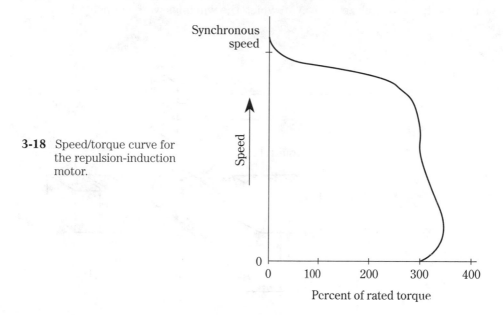

3-18 Speed/torque curve for the repulsion-induction motor.

The polyphase induction motor

The polyphase induction motor is probably the most important type of motor used in industry. The motor is inherently self-starting, and it is very efficient in its conversion of electrical energy to mechanical energy. Polyphase excitation of appropriately designed stator windings produces the *rotating field* that is only simulated by the various devices employed for the starting of single-phase motors. Because this rotating field is symmetrical, torque development is smooth and relatively noiseless.

Although the majority of industrial induction motors are three-phase machines, the primary requisites for achieving polyphase operation are simpler to discuss for a two-phase motor. A basic two-phase induction motor is illustrated in Fig. 3-19. Although four stator windings are symbolically shown, the motor has only *two* poles per phase. In analogous fashion, a two-pole, three-phase motor would have six separate stator windings spaced at 60° intervals. The two-phase induction motor is often encountered in servo systems where the two-phase power can be conveniently generated by solid-state circuitry. Actually, the structure depicted in Fig. 3-19 is virtually the same as that described for the permanent-capacitor induction motor. The chief difference between the two motor types is that the two-phase power source had to be *artificially* produced for the permanent-capacitor motor, whereas such a source is assumed to be available for the two-phase induction motor.

There is nothing unique about the squirrel cage of polyphase machines. Considerable variation in speed regulation and starting torque can be achieved by certain design manipulations involving the resistance of the bars, their number, and the way they are embedded in the armature slots.

Although it is customary to show salient poles in diagrams such as Fig. 3-19, actual induction motors generally have distributed stator windings. It would be rather difficult to determine the number of poles in a stator from visual inspection; you would also have to analyze the way in which the windings were connected.

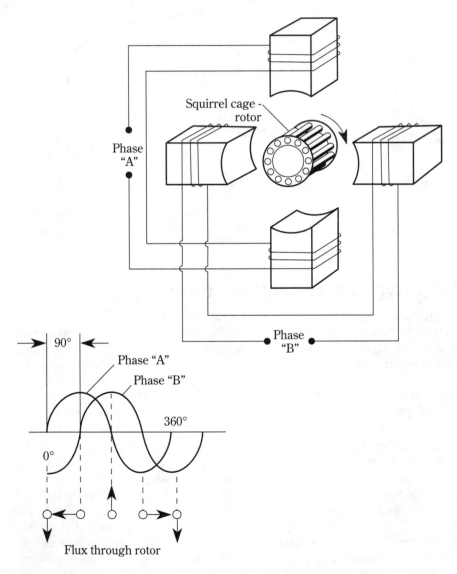

3-19   A two-phase, two-pole squirrel-cage motor—the simplest polyphase machine.

In Fig. 3-19, the flux arrows represent only *instantaneous* polarities—the field being in spatial rotation. Interchanging the motor connection of phases "A" and "B," with respect to the two-phase supply, reverses the direction of rotation. In a three-phase induction motor, the interchange of any two of the three power-line connections will reverse rotation.

A wound-rotor, three-phase induction motor is shown in Fig. 3-20. This drawing is more of a schematic than the pictorial representation shown for the two-phase induction motor. The number of poles in the stator cannot be determined from this drawing; the motor could have any even number of poles. Although the delta connection is shown for the stator, the windings can be designed for connection in the Y configuration, as well. So far, I have not mentioned any difference between this motor and the two-phase machine of Fig. 3-19, except for the different number of phases. If the motor of Fig. 3-20 were depicted with a squirrel-cage rotor, it would be the three-phase version of Fig. 3-19.

The real difference between the motor in Fig. 3-20 and squirrel-cage machine is its wound rotor. Instead of being shorted upon itself, the leads from this rotor are brought out through slip rings. It might appear that this is merely another way of constructing an induction motor. However, the resistance of the wound rotor, and the resistance that might be added with rheostats, develops important operational differences for this motor when compared with the squirrel-cage type.

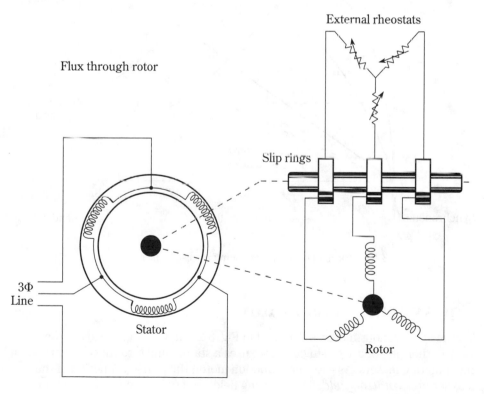

3-20 The wound-rotor, three-phase induction motor.

Inserting resistance in the rotor has two effects. The slip can be considerably increased, thereby providing a method of speed control. Additionally, the starting torque can be made significantly higher. Actually, the maximum torque can be made to occur at 100 percent slip, that is, at standstill, just as the motor is started. This optimum starting situation is realized when the resistance and reactance of each phase are equal. Figure 3-21 shows the control of motor torque by adjustment of rotor resistance. The wound-rotor motor has better starting characteristics and more flexible speed control than its squirrel-cage counterpart. However, it exacts payment for these advantages by having poorer speed regulation and poorer operating efficiency. In Fig. 3-21, the resistance of the rotor circuit is made higher as you progress from curve 1, to curve 2, and to curve 3. If the rotor resistance is *increased* beyond its value in curve 3, it can be seen that the maximum torque capability of the motor is no longer attainable at standstill, or at any other speed. This is the situation shown by curve 4.

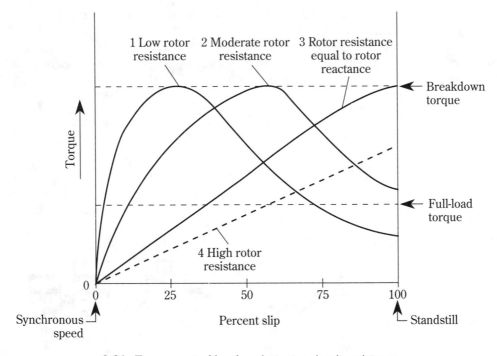

3-21 Torque control by changing rotor-circuit resistance.

The synchronous motor

A basic synchronous motor is depicted in Fig. 3-22. In general, the synchronous motor, whether polyphase or single phase, uses a stator similar to those in corresponding induction motors. As with the induction motor, the primary function of the stator is to set up a *rotating field*. This rotating field can be derived from polyphase windings or from some method such as phase splitting by resistance or by reactance. Now focus attention on the fundamental difference between the two machines.

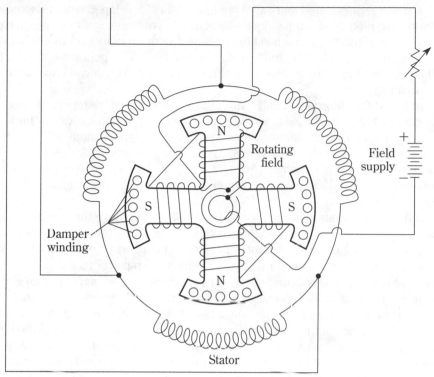

3-22 The three-phase synchronous motor.

As shown in Fig. 3-22, the rotor of the synchronous motor has fixed magnetic poles. These poles can result from the current delivered from a dc source, or they can be incorporated in a permanent-magnet rotor. There is a third possibility—an iron rotor without windings can be used, in which case the required poles develop as the motor pulls into synchronism with the rotating field. In all cases, the rotor is "locked" to the rotating field and can deviate its speed only on a transitory basis. Stated another way, the average slip of an operating synchronous motor is *zero*. Whereas the induction motor *cannot* attain synchronism because it would then develop zero torque, the synchronous motor behaves in an opposite manner. Its torque is developed only at synchronous speed; if forced to slow down by an excessive load, it will first try to fall back a small amount and then try to *resume* the speed of the rotating field. If it cannot develop sufficient torque to do this, it will immediately come to a halt.

The synchronous motor is not a self-starter. Even with the symmetrical rotating field provided by a three-phase stator, *no torque* is developed at standstill. Large synchronous motors are sometimes brought up to synchronous speed by means of another motor coupled to the shaft. This has worked out well in some applications because the other machine could be a dc generator that was operated as a dc motor for starting purposes, then switched over to function as an exciter for the field windings of the synchronous motor. No field current is necessary, or desirable, when the synchronous motor is being accelerated during starting.

A more sophisticated starting technique can be developed from this knowledge of induction motors. By embedding segments of a squirrel-cage structure in the *pole faces* of the rotor, the same machine becomes both an induction motor and a synchronous motor. Such a "hybrid" motor will *start* as an induction motor and then it will attain synchronism because a synchronous motor has a certain amount of "pull-in" capability.

In Fig. 3-22, the squirrel-cage windings are designated "damper windings." This is because of their inhibiting action when the rotor tends to oscillate or "hunt" about its average speed. Even though the average speed is synchronous with the rotating field, the rotor is subject to deviations in its instantaneous speed when there are fluctuations in line voltage or when the mechanical load varies. Another name for these short-circuited conductors in the pole faces is *amortisseur* windings—literally, the "killer" windings.

If the synchronous motor in Fig. 3-22 is driven, it becomes a three-phase alternator. The most familiar example of such an alternator is the one used in automobiles for charging the storage battery. Another interesting characteristic of the synchronous motor is that it is the only ac motor that can control its own power factor. All other ac motors must operate at less than unity power factor, that is, they appear as an *inductive reactance* to the ac line. This causes a higher current consumption than would prevail at unity power factor, and therefore must be considered as a degradation of the overall operating efficiency of the ac motor. By the simple expedient of adjusting the field current in the synchronous motor, the power factor can be varied. The field current can be "resonated" so that the motor appears to the ac line as a resistive load having unity power factor. Even better, the synchronous motor can be operated at a leading power factor to produce an overall unity power factor, taking into consideration other motors and inductive loads. When so operated, the synchronous motor behaves as a capacitor and cancels the inductive reactance of other loads. Indeed, synchronous motors have been made specifically for this purpose; they are called *synchronous condensers* and have no protruding shaft, because they do not drive a mechanical load.

The effects of field current in the synchronous motor are shown in Fig. 3-23. The motor behaves similarly to a resonant circuit because it can be made to have the characteristics of inductance, resistance, or capacitance. Note, also, that its Q is degraded when loaded. These curves indicate the importance of instrumentation in optimizing the performance of the synchronous motor.

The hysteresis synchronous motor

The hysteresis synchronous motor is primarily encountered in fractional horsepower sizes. Its most common application has been in electric clocks and in other timing devices. In larger sizes, and in somewhat modified form, it has also been widely used in tape recorders and phonographs. Figure 3-24 illustrates the clock version, known as the Warren Telechron motor. The stator of this motor is similar to that employed in the shaded-pole induction motor. The rotor, however, is unique. Essentially, it comprises two or more

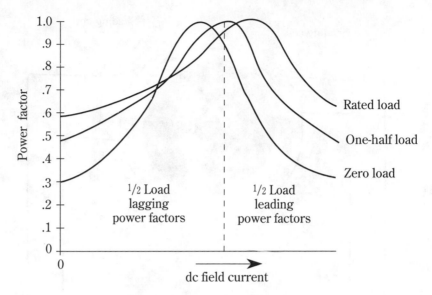

A. Power factor versus field current.

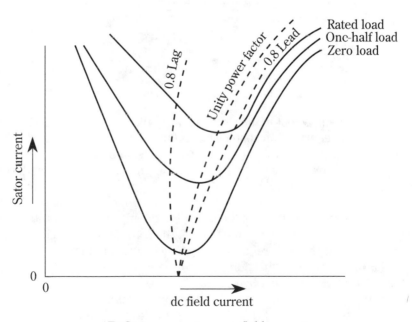

B. Stator current versus field current.

3-23 The effect of field current in the synchronous motor.

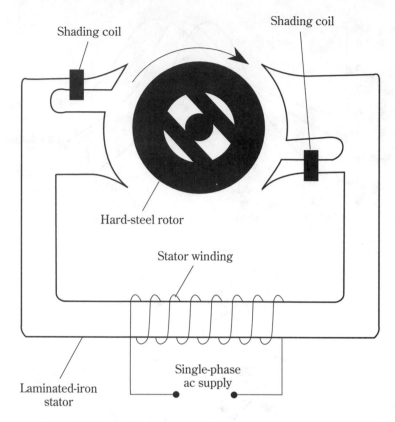

Shading coil

Shading coil

Hard-steel rotor

Stator winding

Single-phase
ac supply

Laminated-iron
stator

3-24 The basic hysteresis synchronous motor.

hard-steel disks stamped out in the pattern shown. Hard steel is used because it is necessary that the material have an appreciable amount of magnetic retentivity. Cobalt steel
has frequently been used as the rotor material for synchronous motors.

When this motor is connected across the ac line, initial torque is developed in
much the same manner as it is in a shaded-pole, squirrel-cage induction motor. The
high resistivity of the steel, together with the dissipational losses caused by magnetic
hysteresis, lowers the efficiency of the motor but greatly increases its starting torque.

Operating in the induction-motor mode, the rotor is brought close to synchronous speed. Unlike the squirrel cage of the shaded-pole induction motor, the steel
rotor of this motor exerts a final burst of torque and attains synchronism with the rotating magnetic field provided by the stator. Once revolving at synchronous speed,
the rotor strongly opposes any tendency to change its speed. This "locking" phenomenon results from the fact that the rotor provides minimum magnetic reluctance
along the length of its cross bars. The rotor develops fixed magnetic poles because
of these cross bars, causing the motor to behave as if it had a simple two-pole bar
magnet spinning within the stator structure. The synchronous speed for 60-Hz excitation is, therefore, 3600 RPM.

Even if the rotor should become completely demagnetized under certain conditions, the motor would still start as described and pull into synchronism. The real importance of the magnetic poles formed in the rotor when it is operating at synchronous speed is the hysteresis produced. As previously stated, such a loss mechanism in the rotor actually increases the amount of torque that is developed. This is important with regard to starting, pulling into synchronism, "locking" at synchronous speed, and turning the external mechanical load.

The reluctance synchronous motor

The reluctance synchronous motor is somewhat similar to the hysteresis synchronous motor—in both machines, start-up and initial acceleration result from induction-motor action. However, the method of implementation in these motors is different. Whereas the hysteresis type lends itself best to small fractional-horsepower motors, the reluctance type is made in much larger sizes, in the integral-horsepower range.

Figure 3-25 illustrates the basic ideas involved in the reluctance synchronous motor. A squirrel-cage rotor with soft-iron segments embedded in its structure is shown in the illustration. The materials and construction of this rotor are so devised that the soft-iron bars, or "windings," can behave as salient magnetic poles. This is achieved by making these bars have a lower magnetic reluctance than any other magnetic material in the armature. This is not the only method by which the desired result can be realized. Any noncylindrical shaping of the armature that results in lower reluctance at certain portions of the periphery than in adjacent portions can produce the required objective. Symmetrically spaced slots filled with nonmagnetic material also can divide the armature into two, or any even number of, regions that can act as magnetic poles.

In any event, the motor starts because the squirrel-cage portion of its rotor experiences a rotating magnetic field. In the example of Fig. 3-25, the rotating field is developed by two-phase windings on the stator. This particular motor must therefore be powered from a two-phase ac source. (Of course, one of the several "phase-splitting" methods can be used if single-phase operation is desired).

When the rotor has been accelerated to perhaps 95 percent of synchronous speed, the iron segments will provide a better path for the flux of the rotating field by being in step with the field. The rotor, therefore, develops an additional increment of torque, which reduces the slip to zero. This condition of synchronism maintains itself because any other speed increases the reluctance distributed to the stator flux. In other words, the rotor "locks in" to the rotating field. During synchronous operation, the squirrel-cage bars develop *zero torque*, but they are not completely out of the picture. Rather, they behave as a damper winding, tending to discourage instantaneous deviation, or "hunting" about the average rotor speed, such as might ensue from line or load disturbances. The motor shown in Fig. 3-25 would have a synchronous speed of 1800 RPM with the stator excited from a two-phase, 60-Hz source.

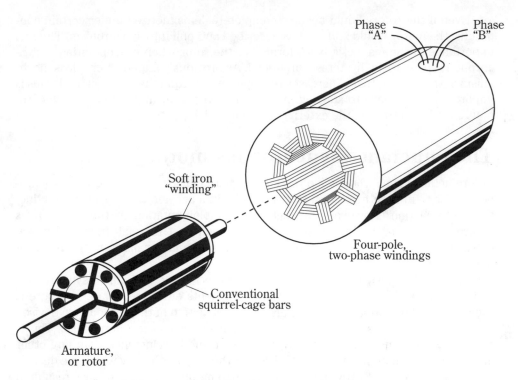

Phase "A"

Phase "B"

Soft iron "winding"

Four-pole, two-phase windings

Conventional squirrel-cage bars

Armature, or rotor

3-25 An example of a reluctance synchronous motor.

The elusive aspect of the induction motor principle

At first glance, the simple experiments depicted in Figs. 1-16C and 1-17 appear to lead to contradictory interpretations. For, in one case, the conducting object obligingly *follows* the moving magnetic field, whereas it eagerly flies away in the other case. This is a dilemma that must be resolved to avoid confusion in understanding real induction motors.

A subtle feature of experiment Fig. 1-16C must be taken into account. Although object X does, indeed, follow the moving magnetic field, it can only do so at a slower speed. This might not be obvious in such a primitively arranged demonstration, for it is conceivable that X might move at 90 or 95 percent of the rate at which the magnet is moved. However, this inherent speed lag is predicated on electrical, not mechanical considerations such as friction or inertia, although such "loading" could produce additional speed retardation.

The important consequence of the above statements is that the relative motion of X is opposite to the direction of the moving magnetic field. Thus, both experiments demonstrate the same phenomenon—there is no contradiction between cause and effect in them. A corollary of these matters can provide further insight: If object X in Fig. 1-16C could attain the same speed as the moving field, the relative motion between the two would be zero, and no eddy currents would be induced in X.

But, without these induced currents, there would also be no associated magnetic field to interact with the field from the moving magnet. This being the case, X would no longer experience any physical force, i.e., "motor action" would cease. Significantly, actual induction motors can approach, but cannot attain synchronous speed (the speed of the rotating magnetic field provided by the stator windings). For sake of simplicity; these matters were not dealt with in chapter 1.

All of the foregoing experiments and allusions can be neatly summed up by Lenz's law, which tells us that an induced magnetic field, such as from eddy currents, must *oppose* the inducing magnetic field. The translation of this fundamental law into the hardware of an induction motor is made easier with the notion of *relative motion*. Notice that X would develop eddy currents, but no lateral motion if the magnet were merely moved up and down in the vertical plane. Similarly, a single-phase induction motor not provided with means to produce a *rotating* electromagnetic field, would not develop any starting torque.

The situation described in Fig. 1-16C also illustrates the principle of the *linear induction motor* used in propulsion of high-speed trains, where relative motion occurs in the continuous horizontal plane, rather than in a circular pattern. Interestingly, the interaction between the inducing and the induced magnetic fields can also be used to levitate the trains so that they have no physical contact with guiding rails.

The asynchronous generator

dc machines are reversible in the sense that when mechanical rotational power is supplied to the shaft, you have an electrical generator. Can this also be said of ac machines? The alternator in the charging system of an automobile could be used as an ac synchronous motor if three-phase ac were applied to its stator windings (with the rectifiers removed or disconnected). The speed of such a motor would be dependent on the frequency of the ac source. Interesting too, this automotive alternator is essentially a miniature version of the large alternators used in utilities power generating stations in which case the speed of the prime mover must be carefully regulated to provide the desired 60-Hz output from the alternator. Such alternators, incidentally, are synchronous alternators—the rotor has fixed magnetic poles usually excited from an auxiliary dc source.

Now focus on induction machines, that is ac machines with either squirrel-cage or wound rotors so that there are no fixed poles. For the sake of simplification, think about single-phase induction motors. Such motors must be specially designed to start inasmuch as there is no starting torque associated with a simple single-phase magnetic force. However, this is not our present problem. Rather the natural question arises as to whether such a machine is also *reversible*—if its shaft is driven faster than its ordinary speed as a motor, does generator action take place?

It is not easy to find a common-sense solution. You might be prone to accept the possibility of generator action, but what about phase relationships? If you already had a source of 60-Hz power and you wanted to send some of it back into the utility 60-Hz line, it wouldn't do to merely make the connection at some random time. It would have to be ascertained that both voltage and phase were correct. Such synchronization must be performed very carefully to prevent blown fuses, activated circuit breakers, or worse. How then can an induction motor be expected to smoothly and obligingly transform itself from a motor to a generator if its shaft speed is speeded up from an external source of mechanical power? Surely, it can be argued that either there would be no generator action, or that there would be fireworks if generator action did take place.

This is not merely an academic probe; the nature of such hypothetical operation has some very practical ramifications. If generator action takes place, regenerative braking should be possible on electric vehicles. Also, some utilities will pay for power injected *into* the ac line. It could be much simpler to use a speeded-up inductive motor as the generator than to attempt the use of a synchronous alternator by the environmentalist interested in tapping alternative sources of energy, such as a fast-flowing stream, or the wind.

The actual behavior of an induction motor as a function of its shaft speed is shown in Fig. 3-26. It is helpful to associate these curves with a physical system such as a train powered by an inductive motor. At standstill and while accelerating to its running speed, high mechanical power is required from the shaft and high current is consumed by the motor to meet this demand. The motor ultimately attains its running speed. This is somewhat *below* synchronous speed—the actual rate of rotation of the magnetic field supplied by the stator windings. The motor cannot *attain* synchronous speed even at no load, because there would then be no electromagnetic torque to cause the rotor to rotate.

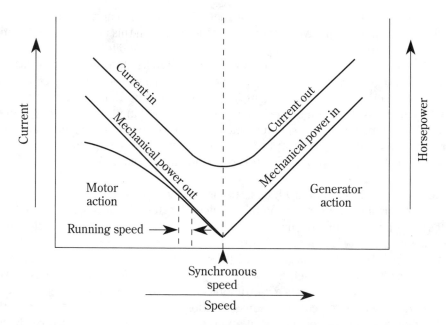

3-26 An induction motor becomes an induction generator above-synchronous speed. If appropriate mechanical power is imparted to the shaft of a line-connected induction motor, higher than synchronous speed will cause current to be injected *into* the line. The motor will have become a generator.

Now, suppose the train encounters a down-hill grade. This can result in additional mechanical motion imparted to the shaft, causing the motor to run at higher-than-synchronous speed. The motor now behaves as a *generator*, sending current back into the ac line. This appears reasonable enough, but it is only natural to ponder the effect of the frequency of the generated current. It fortunately turns out that the generated frequency will always be 60 Hz, or whatever the ac line frequency is. Moreover, this is true no matter how much the shaft speed of the machine exceeds synchronous speed. This wonderful situation exists because the magnetic field in the rotor is induced from the current in the stator windings, rather than being the result of a permanent magnet, or of dc from an external source. Such operation justifies the name of the machine as an *asynchronous generator*, it is essentially a mechanically driven induction motor.

The curves in Fig. 3-26 show that at synchronous speed, the practical induction motor still consumes some current and still requires some mechanical power. This is because the magnetizing current and the electrical and mechanical losses are not zero as they might be postulated to be in an ideal machine. Also, as pointed out, no torque is developed at synchronous speed so an external source of mechanical power is needed to turn the shaft. Neither the practical nor the ideal induction motor can still behave as a motor at exact synchronous speed. And, unlike the synchronous generator, the *phase* of the current presented to the ac line by the asynchronous generator is always automatically correct.

Electrified railroads have, indeed, made good use of the induction motor in the manner alluded to. Because of its rugged construction that dispenses with the high current-carrying brushes of dc motors, it was long hoped that it could be used in electric automobiles. This, however, had to await the development of efficient and cost effective solid-state inverters. In the near future, it is likely that there will be considerable competition between the use of ac and dc motors for electric vehicles.

A shortcoming of the inductive generator is that it does not like to deliver current into a line or load displaying a lagging power factor. Static or synchronous condensers might be used to alleviate this problem, but such remedies negate the features of easy implementation and low cost.

Regenerative braking

In our discussion of the asynchronous generator (actually, an induction motor caused to run *above* its synchronous speed) a simple illustrative example was postulated of a train fed from an ac line and propelled by an induction motor. On a downhill grade, the motor could be caused to operate above its synchronous speed, thus becoming an asynchronous generator delivering power back *into* the ac line. This kind of operation can also manifest itself as regenerative braking—slowing down the train and relieving wear on the frictional brakes. Note that energy is returned to the ac power line rather than being dissipated in a resistance or in the motor itself as in *dynamic* braking.

However, such a simplified propulsion system with the induction motor powered directly from the ac line cannot provide much braking effort—most of the time, the motor would not be acting as an asynchronous generator and would be busy adding to, rather than subtracting from, the tractive effort of the train. In practice, therefore, a different arrangement is used—one that permits regenerative braking almost down to zero speed of the train. Such action greatly enhances safety, decreases maintenance of the frictional brakes, and increases the overall operating efficiency of the electrification system.

Such a system is shown in Fig. 3-27. Notice that the input of electrical power can be from either an ac or a dc line. If an ac line is involved, the line conditioner is a rectifier, if operation is from a dc line, the line conditioner would be a filter to protect railroad signaling apparatus from noise and transient contamination. The basic idea in either case is to have dc power available to operate a *variable frequency inverter*. Such an inverter is often designed around GTOs with ratings at kilovolt and kiloampere levels. By supplying the gates of the GTOs with appropriate control pulses, the induction motor can be operated from a variable-frequency, three-phase format. Additionally, the GTO gate logic makes use of pulse-width modulation to increase the voltage applied to the motor as frequency is increased.

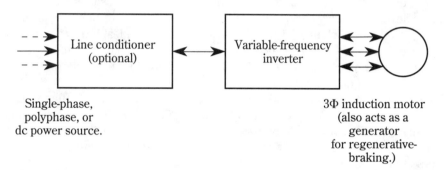

Single-phase, polyphase, or dc power source.

3Φ induction motor (also acts as a generator for regenerative-braking.)

3-27   Basic scheme for regenerative braking from induction-motor drive. The *frequency* of the inverter is controlled by the vehicle operator. When the frequency is low enough while the motor is still turning at above-synchronous speed, the motor becomes a generator, sending power back into the power source. This causes regenerative braking.

With such an arrangement, there remains one more requisite if regenerative braking is to take place. The inverter and the rectifier, if one is used, must allow passage of current from load to line as well as from line to load. This might sound like a contradiction of function, but no polarity reversal of dc voltage is involved. Although such a bidirectional current path does not exist in diode circuits, thyristors (such as SCRs and GTOs) can allow bidirectional current. Accordingly, when the induction motor begins to behave as an asynchronous generator, power flows back to the line and *regenerative braking* takes place. The thyristors are said to operate in both the first and second quadrant of their conduction characteristics. Regenerative braking is brought into play when the train operator lowers the frequency of the inverter so that the motor is turning faster than its new synchronous speed.

Concatenation of induction motors

The concatenation principle of operating induction motors is best described in older texts on electrical engineering. The actual word is not a technical term, per se, but simply demotes things or events arranged in a *series*. Concatenation pertains to an old technique of connecting induction motors. It is relevant to know that the technique was largely used with traction applications of electric motors, especially the railroads. It might soon enough assume relevancy again as interest in electric vehicles as alternative to cars propelled by gasoline and diesel engines increases.

The basic idea underlying concatenation is to enable more than one speed to be obtained when using the induction motor(s) for vehicle propulsion. This can be accomplished by changing the frequency, changing the number of poles, or by causing the slip between the rotor and the rotating electromagnetic field from the stator to change. Until recently, there was no practical way to change the frequency. Changing the number of poles is workable for ceiling fans, but tends to seriously compromise efficiency when used in a traction vehicle. Inasmuch as the immediate source of energy in electric vehicles is the battery, motor efficiency is of primary importance if acceptable range between charging is to be realized. Causing rotor slip can be advantageously used, but not by simply loading down the shaft of a single induction motor.

In concatenation, primary reliance for speed change results from a unique use of change in *slip* and the resultant change in *frequency*. This is brought about by the use of two induction motors coupled to the same shaft. At least one of these induction motors must be of the wound-rotor type. The other can be a squirrel-cage induction motor. In order for the set to be self-starting, polyphase motors have generally been employed in this scheme. The basic arrangement is shown in Fig. 3-28. The simplest use of this technique makes use of two motors with identical numbers of poles. This, however, is not a requisite—motors with unequal poles can be used to yield a more desirable selection of speeds. An added feature of concatenation is the ease of reversing rotation by electrical switching of the phase connections of the ac line, or of the machine interconnections.

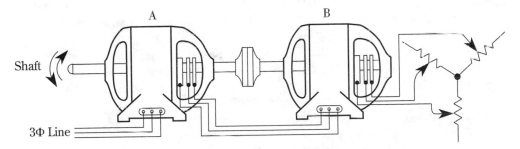

3-28 Basic setup for concatenation of induction motors. By appropriate selection of the interconnections, it is possible to obtain either three forward and one-reverse speed, or three reverse and one forward-speed. Otherwise, actual speed governed by the number of poles in the motors and by the line frequency.

In the ensuing discussion of this technique, keep in mind that during its heyday, there were no solid-state frequency changers such as cycloconverters or three-phase inverters. Rather than relegate concatenation into obsolescence, it is likely that the newer solid-state circuits can be applied in mutuality to produce enhanced performance and extended operational flexibility to electric-vehicle propulsion.

In determining what speeds are available from a setup such as that of Fig. 3-28, it is convenient to deal with synchronous speeds. These will be approximate in the sense that induction motors do not actually run at synchronous speed, but always slightly lower. Thus, a four-pole induction motor operating from a 60-Hz source does not rotate the synchronous speed of 1800 RPM, but rather at a slower rate, say 1750 RPM. Synchronous speed is the rotational rate of the electromagnetic field set up by the stator windings. Ideally, the mechanical rotation of the rotor should also be at this synchronous speed. However, it is the nature of the induction motor that zero torque would be developed if the rotor actually *attained* synchronous speed. Because the operating speed is quite close to the synchronous speed, it is not uncommon to identify a 60-Hz four-pole induction motor as an 1800-RPM motor, a 2-pole induction motor operating from 60 Hz as a 3600-RPM motor, etc.

For purposes of more exact determination of speed, the synchronous speed and the rotor rotation speed give rise to a third quantity, the *slip*. Slip is expressed as a percentage, and is established as follows:

$$s = \frac{100\ (synchronous\ speed - rotor\ speed)}{synchronous\ speed}$$

At standstill, slip is 100 percent and the frequency of the current induced in the rotor is the same as the ac supply frequency. If the rotor could attain synchronous speed, its induced frequency would be zero. Inasmuch as the slip in the conventionally operating induction motor is low, say two or three percent, the frequency of the current induced in the rotor is relatively low. Thus, if a 60-Hz induction motor operates with a slip of 2.5 percent, the frequency of the rotor current would only be 2.5 percent × 60 or 1.5 Hz. And, during startup and acceleration, the rotor frequency would change progressively from 60 Hz to 1.5 Hz.

Contemplation of the above facts reveals the induction motor as an inherent *frequency changer* with an "internal" frequency quite different from the ac frequency applied to the stator windings! The idea naturally presents itself that a second induction motor powered from the current induced in the rotor of the first motor would tend to run at another (lower) speed. Indeed, the two motors depicted in this "series" or concatenated relationship in Fig. 3-28 must exhibit unusual speed characteristics. Inasmuch as the two motors share a common shaft, it is not easy, by intuition alone, to ascertain the shaft speed(s) of this interesting arrangement.

It turns out that, to a good practical approximation, the shaft speed of such a motor combination is equal to that of a fictitious third motor having either the sum or difference of poles in the actual motors. Generally, this allows the choice of four different speeds, with one choice being a reverse-rotation speed. How might this work out in an example? Assume the ac line frequency is 60 Hz.

Let one of the two shaft-coupled induction motors have four poles, while its companion has eight poles. The synchronous speed of the four-pole motor is 1800

RPM. The synchronous speed of the eight-pole motor is 900 RPM. These, then, are two of the four approximate no-lead speeds that are available. The speeds are said to be approximate in the sense that induction motors operate close to, but not at, synchronous speed. Additionally, an approximate speed will be available corresponding to a twelve-pole induction motor (the "fictitious" motor alluded to). Such a twelve-pole motor would have a synchronous speed of 600 RPM.

Yet, *another* speed is available by connecting the stators so that the motors tend to rotate in opposite directions. Under this condition, you have a four-pole "fictitious" motor turning at 1800 RPM in the *reverse* direction. Summing up, by merely changing the stator connections you obtain at least four shaft speeds. ("At least" four, because the first three speeds and the fourth speed could also have been traded, making eight speed situation in all).

It can be appreciated that, depending on the selection of the motors, a variety of speeds can be provided. Also, with modern variable-frequency solid-state inverters, one isn't restricted to 60 Hz and the speed range of the combination can be made continuous over a wide range. Also, motor "B" in Fig. 3-28 can be a simpler squirrel-cage induction motor rather than the wound-rotor type shown. The possible advantage of the wound-rotor machine over the squirrel-cage type lies in its higher starting torque.

In the event this technique is adopted for electric vehicles, it is likely that the two induction motors will be constructed as a single machine. One purpose of its use will be to eliminate need for a mechanical transmission.

Incidentally, in those speed situations in which the resultant shaft speed is due to the two motors tending to turn in opposite directions, the concatenated motors are never self-starting. The shaft must *already* be turning in the desired direction, even for a brief instant. This can readily be taken care of by the switching arrangement in the motor leads.

The ac powered shunt motor—Where is it?

An interesting question is why we don't see dc shunt motors operated as ac machines? This is a natural question in light of the fact that the dc series motor (or optimized versions thereof) can yield very satisfactory performance when powered from an ac source. It would appear that the same logic would pertain to both machines—the field and armature undergo simultaneous polarity reversals and the resultant torque should be unidirectional with both machines. That is, both should develop net motor action on ac.

If you encounter this dilemma for the first time you might suspect a fallacy in the logic as worded. The fallacy has to do with the word, "simultaneous". It might appear that relativity theory is being brought into the discussion. However, the use of the word here has a simple implication. Although the same *voltage* is simultaneously impressed across both armature and field terminals, their currents do not both simultaneously rise and fall with the applied ac voltage. Because of the large inductance of the shunt field winding, the field current is out of phase with the armature current. Unfortunately, this phase difference amounts to nearly 90° in practical shunt motors. This quadrature lag in the phase of the field current is shown in Fig. 3-29.

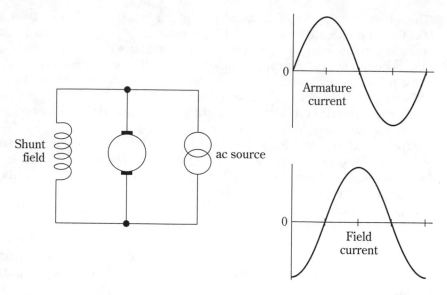

3-29   Unlike the series motor, the dc shunt motor is not suited for ac operation. The high inductance of the shunt field causes armature and shunt field currents to be very much out of phase. There is little opportunity for torque-producing interaction between their magnetic fields.

Inasmuch as the magnetic field of the shunt field winding is produced by its ampere turns, it follows that a lagging current is accompanied by a time-delayed magnetic field. This tells us that when the magnetic field of the armature is maximum, the magnetic field from the field winding is at, or near, minimum. The converse relationship is also true for these magnetic fields. Inasmuch as motor action depends on the *interaction* between these two magnetic fields, it is clear that the quadrature phase relationship allows little opportunity for such interaction. Therefore, the ac-powered shunt motor either does not run at all, or develops impractically small torque. Contrast this situation to that of the series motor where the *common* current through the field winding and the armature allows interaction between *simultaneously* occurring magnetic fields.

While on the subject, it should be obvious why a permanent-magnet dc motor cannot run on ac. Here, to be sure, there is strong interaction between the motor fields. The trouble is that this interaction is at one instant clockwise, then counterclockwise at the next instant, and so on. The net result is that the motor is free to vibrate, but not to develop torque in either direction. And unlike certain single-phase ac motors, no amount of spin of the shaft can provoke the permanent-magnet dc motor to operate on ac power.

4

Electronic control of commutator-type machines

THE STUDY OF ELECTRIC MOTORS AND GENERATORS GENERALLY BEGINS WITH DC machines and progresses to ac types. This, like many classifications, is quite arbitrary. Indeed, valid arguments can be readily offered for the *reverse* treatment of the subject matter. For example, it is true that the armatures of virtually all dc machines actually carry *alternating current*. The reason that certain machines are said to be dc types is because they have a commutator. The commutator is a *rectifier*, or a reversing switch, which adds complexity to a machine. Thus, it would not be unreasonable to study ac machines first and then progress to dc motors and generators.

For the sake of convenience, a classification plan can be adopted for the study of electronic-control applications to electrical machines. Rather than deal with the concept of dc and ac motors and generators, these machines can be treated on the basis of whether or not they have commutators. This is my method of covering circuits and systems discussed in this chapter and in the next chapter. I will thereby circumvent the dilemma of deciding the most appropriate way to classify solid-state controllers of *universal motors*.

This chapter investigates the electronic control of machines with commutators. The fact that some circuits or systems are intended for operation from dc sources, while others are powered from the ac line, is certainly a point of interest. The primary classification criterion will, however, involve the presence of the commutator.

Speed control of a universal motor with SCR

The salient feature of the speed control shown in Fig. 4-1 is *simplicity*. It is a half-wave system because the single silicon controlled rectifier allows passage of alternate pulses from the power line. These unidirectional pulses are fractional parts of the alternations—how large a fraction is governed by the time in the ac cycle that the SCR is triggered. This timing is, in turn, determined by the phase shift produced at the junction of C1 and P1 with respect to the cathode of the SCR. Control of this phase shift is provided by adjustment of P1.

Although the motor receives pulsed energy by means of this circuit action, the repetition rate of the pulses is fast enough to develop an essentially smooth torque.

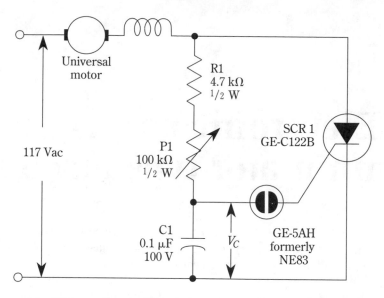

4-1 SCR speed control circuit for universal motors. <small>General Electric Co.</small>

The later the SCR fires in the ac cycle, the less the average current is through the motor. Accordingly, less internal torque is developed and the speed is thereby decreased. In such a half-wave system, the top speed of the motor is less than what it would be if it were connected directly across the power line. This is readily obvious from the following consideration. Top speed would be obtained with the SCR conducting a full 180° of each ac alternation. But, an SCR operating in this fashion would simulate an ordinary rectifier diode. The motor would then "see" a simple half-wave power source. The half-wave rectification makes 45 percent of the average current available to a load that would be available with a direct connection (the concept of average current rather than RMS values is relevant to motor operation because torque depends on *average* current). In some applications, a switch is placed across the SCR in order to provide the full-power option for motor operation. Another limiting factor with regard to control range is the inability of the phase-shift circuit to provide the ideal 0- to 180-degree span adjustment.

Even with the low rectification efficiency of half-wave operation, the process occurs with very little energy dissipation. Therefore, motor control by a circuit such as that of Fig. 4-1 is much more efficient than simple rheostat speed control. Also, the basic speed/torque characteristic of the motor is not appreciably altered. The only important dissipation in the SCR is that associated with its forward voltage drop, which is relatively low. The overall result is that large motors can be controlled in this fashion with minimal concern for heat removal.

The neon bulb, because of its high firing voltage, is relatively immune to erratic triggering. This sometimes is an important consideration due to the transients imposed on the line by brush sparking. Although commonplace neon bulbs might suffice for use with very small motors, it is generally best to use the specially designed units. This is because high current pulses are needed to fire SCRs.

Regulated speed control of universal motors

The speed-control circuit of Fig. 4-2 has much more to offer than might be apparent from casual inspection. Although no obvious connection between output and input is discernible, this half-wave circuit nevertheless incorporates *feedback*. Because of this, the speed of the universal motor at any setting of potentiometer P1 does not deviate greatly over a wide torque range. Thus, the "natural" characteristics of the universal motor are *electronically modified*. This is highly significant, because it exemplifies a unique advantage of electronic control; it enables the motor designer to produce a motor with optimized cost, commutation, and flexibility. The actual speed/torque behavior then can be manipulated electronically. In this case, the motor with the *poorest* speed regulation can be converted to near constant speed performance.

Feedback occurs in the following manner: When the SCR is not conducting, the rotating motor continues to generate a counter EMF, which is polarized to inhibit triggering of the SCR. In order for the SCR to trigger, a voltage that is equal to its triggering voltage plus the counter EMF of the motor must be applied to its gate circuit. It receives this voltage from the low-impedance output of emitter-follower stage Q1. Stage Q1, in turn, samples the triggering voltage from the divider network comprising R1, P1, CR2, C1, and C2. Transistor Q1 functions as a current amplifier and is instrumental in extending the low-speed range of the motor where excessive loading of the divider network tends to be detrimental. At any given setting of potentiometer P1, the voltage applied to the SCR gate by Q1 can be considered a

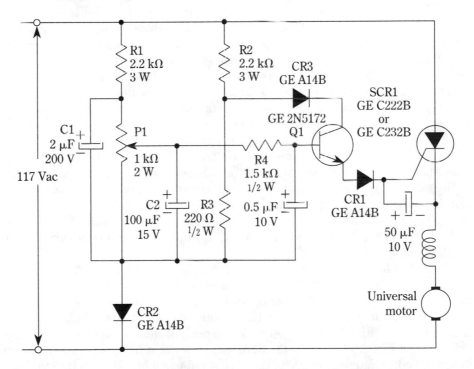

4-2 Feedback-regulated SCR speed control for universal motor. General Electric Co.

reference voltage. The SCR gate circuit "compares" this reference voltage with the counter EMF of the motor. Suppose that the motor is operating and additional mechanical load is imposed on its shaft. In the manner of series machines, it will attempt to slow down drastically and consume more torque-producing current. However, such a slowdown is accompanied by decreased counter EMF. This, in turn, enables the SCR to conduct with a lower output voltage from Q1. Such a lower voltage is available earlier in the ac cycle, so the SCR now delivers an increased average voltage to the motor. The motor develops increased torque, which accelerates it back to the vicinity of its previous speed.

In the event that the load on the motor is relaxed, the attempt of the motor to speed up is counteracted by a sequence of events opposite to those described above. Because of the electronic control, the universal motor is forced to behave in a manner similar to that of the dc shunt motor. The latter machine is often referred to as a "constant-speed" motor.

Triac speed control for universal motors

The circuit shown in Fig. 4-3 features a wide speed-control range, together with smooth motor performance. These desirable characteristics stem from the use of full-wave triac control and a double phase-shift network. The full-wave control enables application of nearly full power to the motor at the high end of the control range. As previously pointed out, half-wave control is a rectification process; a half-wave rectifier working over the full 180° of the ac half cycle can deliver only 45 percent of the average load current that would be available directly from the ac line. Another disadvantage of half-wave control arises when it is desired to use an isolation transformer in the incoming line—the dc component that accompanies half-wave rectification tends to saturate the transformer, thereby increasing its losses and distorting the voltage waveshape in the secondary winding.

The triac does *not* provide full-wave rectification; its control mechanism is even better for the purpose at hand. The triac functions as would two SCRs connected back to back, but without the gating complications attending the use of dual SCRs. The voltage delivered to the load is always an ac sine wave. Variable fractions of a true sine wave are produced by varying the timing of the gate trigger signal. Because of the aforementioned wave symmetry, there is no dc component. Because rectification is not involved, a universal motor would be expected to perform much better than a dc series motor. Also, this circuit could be used to provide limited speed control to certain types of induction motors, such as the permanent-capacitor and shaded-pole types.

The double phase-shift network comprises R1, C1, R2, and C2. This network practically eliminates a disturbing hysteresis effect which characterizes single phase-shift networks when a thyristor is operated at low conduction angles—this corresponds to low motor speeds. For lamp loads, the hysteresis effect is merely a nuisance; you learn to exercise patience when adjusting for dim illumination. However, for motor control, the effect is more serious because erratic operation will occur at low speeds. Hysteresis is due to the abrupt reduction of voltage across timing capacitor C1 when triggering occurs. Techniques for overcoming this effect involve imposing less loading on this capacitor.

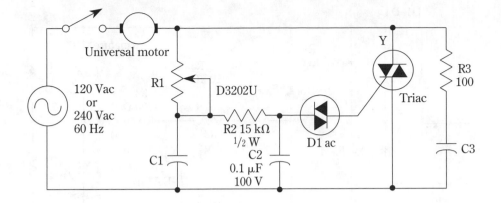

4-3 Triac speed-control for universal motors. General Electric Co.

	120 Vac, 60 Hz	240 Vac, 60 Hz
R1	100 kΩ, $1/2$ W	200 kΩ, 1W
C1	0.1 μF, 200 V	0.1 μF, 400 V
C3	0.22 μF, 200 V	0.22 μF, 400 V
Y	T2700B	T2700D

The "snubber" network, R3 and C3, connected across the triac is a necessary precaution when inductive loads, such as motors, are used. Otherwise, the inductive kickback pulse developed when the triac switches to its off state can have a rate of rise that is sufficient to retrigger the triac regardless of gating conditions.

Full-wave speed control for shunt motors

The speed control circuit of Fig. 4-4 can provide a wide range of speed adjustments. At the same time, the speed regulation is generally superior to that of the inherent behavior of the motor itself. This controller was specifically developed for use with dc shunt motors. Speed torque curves obtained with a ⅛-hp dc shunt motor are shown in Fig. 4-5.

Unlike many thyristor motor-control schemes, this one converts the power from the ac line to dc via a bridge rectifier (D1 through D4). The motor field is permanently connected across the dc output of this bridge. The armature of the motor receives variable duty-cycle current pulses from the SCR, which interrupts the full-wave dc from the bridge rectifier. It might initially appear that there would be a commutation problem with the SCR, as is usually the case when dc is chopped or interrupted. However, this is not the case with the circuit in Fig. 4-4. The unidirectional current obtained from the rectifying bridge dips to zero twice each cycle,

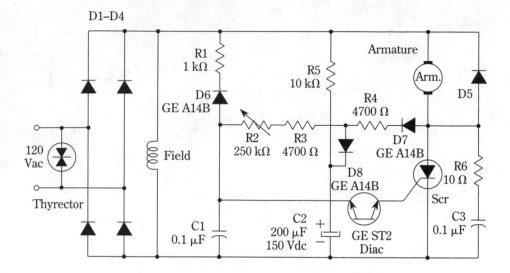

Note: Motor size determines types SCR, D1–D5 and thyrector

4-4 Speed control for dc shunt motor. General Electric Co.

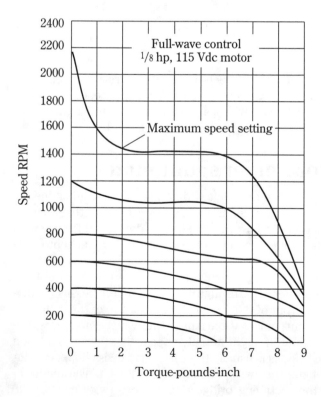

4-5 Typical speed/torque curves obtained with the full-wave control circuit. General Electric Co.

because it is not filtered. This being the case, the SCR extinguishes its conduction twice for each full cycle of the incoming ac power. From a 60-Hz line, the SCR can be triggered into its on state 120 times per second, rather than 60, as in conventional half-wave circuits. Accordingly, it must be classified as a full-wave controller.

In other respects, this SCR circuit operates similarly to many other motor-speed controllers. The time that it takes timing capacitor C1 to charge to the breakdown voltage of the diac determines the triggering time of the SCR. If the SCR is triggered early in the excursion of a half-cycle of the rectified waveform, a *large* amount of average dc power will be delivered to the motor armature. If, however, speed adjustment control R2 is set to slow down the charging rate of capacitor C1, triggering voltage will be reached at a later time. Therefore, a relatively small amount of average dc power will flow into the armature and the motor speed will be reduced.

Although a feedback path might not be obvious in the schematic diagram, negative feedback is a major feature of this arrangement. Therefore, the motor speed not only is adjustable, but can be *regulated* as well. Suppose that the motor is suddenly loaded and attempts to drop its speed considerably. Such a slowdown would be accompanied by a reduction in the counter EMF, thereby enabling timing capacitor C1 to charge faster through D7, R4, R3, and R2. Triggering voltage is then developed across C1 *earlier* in the half cycle, and the motor armature receives a higher average current, thereby accelerating it to counteract the drop in speed. One should be mindful that the voltage monitored by D7 is the anode voltage of the SCR and is the output of the bridge rectifier minus the armature counter EMF. The opposite reactions occur if the motor attempts to speed up. You can now identify the feedback path as comprising D7, R4, R3, and R2.

Feedback speed-control circuit for permanent-magnet motors

The inherent droop in the speed torque characteristics curve of the permanent-magnet motor can be largely overcome by means of a feedback arrangement such as that depicted in the schematic diagram of Fig. 4-6 (this scheme is also applicable to shunt motors, in which case it would probably be preferable to excite the field from a constant-current source). The improvement obtainable with this technique is shown in the curves of Fig. 4-7. It is relevant to point out that the downward slopes of the curves in Fig. 4-7A would be even steeper if rheostatic control of armature current were used.

This is an "on-off" system in which operating current is parcelled out to the motor armature in such a way as to maintain a near-constant speed. It bears a resemblance to pulse-width modulation but is not ordinarily so designated because both the on time and the current vary. It shares the salient feature of pulse-width modulation, however, in that motor current is either on or off—there is no power dissipation from intermediate current values as in rheostatic control. The average value of armature current is the controlled parameter, and this is a function of the average duty cycle.

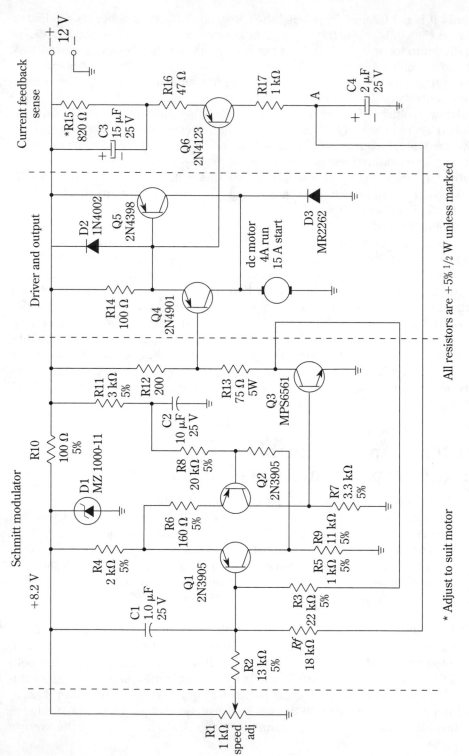

4-6 dc motor speed-control circuit utilizing current feedback. Motorola Semiconductor Products, Inc.

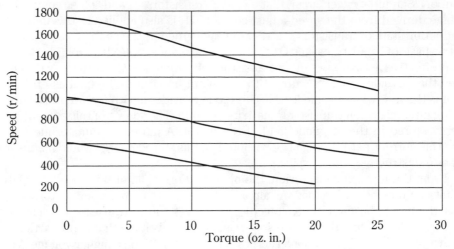

A. "Natural" characteristics when motor is powered from a tapped battery or a voltage - regulated power supply.

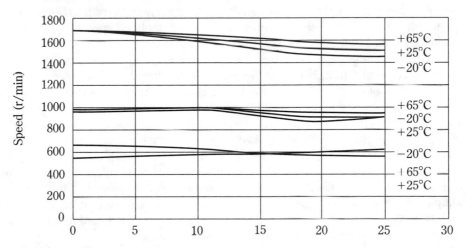

B. Speed-torque characteristics when the same motor is controlled by the feedback circuit in Fig. 4-6.

4-7 Speed-regulation characteristics of a permanent-magnet dc motor. Motorola Semiconductor Products, Inc.

The sensed parameter is not directly motor speed, rather, it is the voltage drop across the base-emitter junction of power-output transistor Q5. This quantity, V_{BE}, varies with motor *current*. For example, if increased mechanical load is applied to the motor, it tends to slow down and decrease its counter EMF so that it can consume more torque-producing current. Voltage drop V_{BE} in Q5 then becomes greater because of higher collector current. The sense transistor is Q6, and the dc voltage that *represents* motor speed is developed at point "A." This voltage is applied to the input of a Schmitt-trigger circuit that includes transistors Q1 and Q2. A "comparison" voltage derived from the speed-adjust control, R1, is also applied to this input. The conduction state of transistor Q2 is therefore governed by the net dc voltage sampled by the base of input transistor Q1. Transistor Q3 is a simple dc amplifier stage for more effective actuation of drive transistor Q4. Transistor Q5 is the power-output stage that actually "meters" current to the motor armature. The final essential element is the free-wheeling diode, D3. This diode provides a path for armature current during intervals when transistor Q5 is in its off state. The source of this current is the energy stored in the magnetic field of the motor. A more-constant torque results from this current path through D3. Without D3, this stored energy would manifest itself as a destructive voltage spike.

From the foregoing discussion, the operating principle of this speed-regulating circuit can now be understood. As already pointed out, an increased motor load results in a higher base-emitter voltage, V_{BE}, in power-output transistor Q5. This voltage is sampled by feedback-sense stage Q6, and a dc voltage that is proportional to the average V_{BE} in Q5 is developed across capacitor C4. When a higher voltage is thus developed across capacitor C4, it causes input stage Q1 of the Schmitt trigger to *shorten* its conduction intervals. The resultant *increase* in on time for stage Q2, and for all subsequent stages through Q5, causes a higher average current to be delivered to the motor armature, thereby restoring much of the depleted speed. The opposite reactions take place when the load on the motor is relaxed.

A little contemplation reveals that the circuit of Fig. 4-6 involves positive feedback (an increase in armature current begets a further increase). However, the feedback factor is not sufficient to cause any instability. The motor itself tends to discourage a cumulative buildup to either a "hung-up" or an oscillatory condition because as each increment of additional current produces increased speed, the counter EMF is increased, thereby inhibiting development of runaway current.

The Schmitt-trigger voltage divider, made up of R5, R8, R9, and R11, connects across the 12-volt power source. Because of this, if the voltage from the power source becomes higher, transistor Q2 will be biased on for a shorter time, thereby tending to prevent increased average current from being delivered to the motor. Although this compensating action is not perfect, it makes motor speed much less dependent on supply voltage than would otherwise be the case.

Because of varying characteristics in motors and wide tolerances in transistor parameters, the optimum value for resistance R15 will generally require trial-and-error determination.

Independent speed and torque control of shunt motors

In the shunt motor, and the permanent-magnet motor as well, the electromagnetic torque developed in the armature is almost directly proportional to armature current. Therefore, limiting the armature current also defines the maximum torque that can be developed. This follows from the equation for torque T, where $T = k\,I_a\,\Phi$. In this relationship, constant k is a function of the design parameters of a given motor, such as the number of poles, the characteristics of the magnetic material, the dimensions of the air gap, etc. Because field strength Φ is, or can be, nearly constant in a shunt motor or a permanent-magnet motor, it follows that torque is directly controllable by means of armature current. Of course, the control of *speed* is generally assumed to be derived from torque control and the two performance parameters might appear to be interdependent.

However, the speed of a shunt motor or permanent-magnet motor is fairly constant with respect to the torque extracted from the shaft. Moreover, the speed of these motors can actually be controlled by armature voltage while holding armature current constant. What is needed is a controller that will produce abrupt *current limiting*, as is attainable from laboratory power supplies that have current-limiting controls.

The setup shown in Fig. 4-8 allows a dc permanent-magnet or shunt motor to be independently controlled in either speed or torque. The technique is straightforward with a linear power supply. Greater overall efficiency can be realized with a regulated supply using thyristors or premised on the switchmode principle. The voltage and current controls on the regulated supply must be set high enough to enable the motor to operate at the maximum desired speed and torque. Inasmuch as these motors naturally have fairly flat speed versus load curves, it is possible to dispense with the tachometer voltage-feedback loop in some practical cases. With both feedback loops operative, the speed vs. torque characteristics will be much as depicted in the curves of Fig. 4-9. Here, torque depends on the level of constant current delivered to the armature, speed depends on the level of constant voltage applied to the armature.

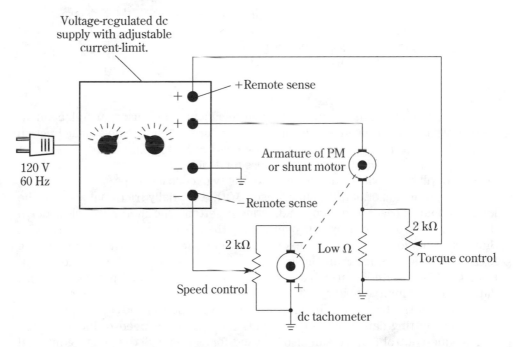

4-8 Independent speed and torque control/regulation of pm or shunt dc motor. Speed is primarily a function of armature *voltage*. Torque is primarily a function of armature *current*. The voltage-regulated dc supply automatically becomes a current regulator when a preset level of armature current is reached.

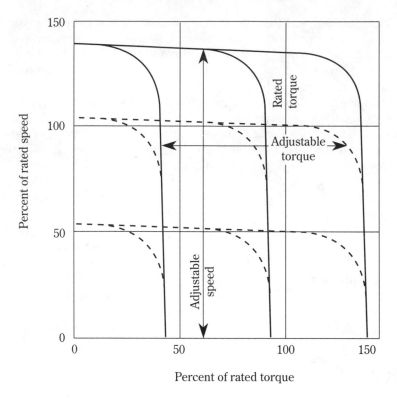

4-9 Curves showing independent speed and torque control of a pm or shunt dc motor. Bodine Electric Co.

If a shunt motor rather than a permanent-magnet motor is used, an independent field current source is needed. Inasmuch as field current tends to be a small fraction of armature current, minimal cost and complexity are involved. Indeed, it need not even be regulated in most instances because tendencies for small variations in speed or torque will be stabilized via the action of the two feedback loops.

The torque adjustment has many uses. With essentially frictional loads, if the load demand exceeds the torque adjustment setting, the motor will slow down drastically, or stop. This prevents overloading the horsepower rating of the motor. Similarly, the motor will not overexert itself when presented with loads requiring higher starting torques than those allowed by the setting of the torque poten-tiometer. These safety features are valuable in many industrial processes where damage and/or injury could result from the attempt of a drive motor to overcome a "jam-up" in the driven machinery. Gearmotors, in particular, can be protected from transmitting dangerous torque levels through their speed-reduction units. The torque-control feature can also be used for "softening" the starting interval and for controlling acceleration to a preset speed when essentially inertial loads are used.

The torque control of dc motors

As already shown, speed is not the only motor operating parameter amendable to electronic control. Others are starting, stopping, reversal, positioning, torque, and horsepower. Another interesting and useful control technique is shown in Fig. 4-10. Here, again, the *torque* of a dc motor (usually a permanent-magnet or a shunt type) is controlled independently of speed. This is achieved with *current feedback*. The scheme is analogous to that employed in current-regulated power supplies. The torque command voltage is, in essence, an adjustable reference source. For any given value of torque-command voltage, the current in the armature of the motor is maintained at a constant value. Because the torque is proportional to armature current, it also is "programmed" by the torque-command voltage.

The motor is by its nature a current-to-torque transducer. The power converter is, because of the current feedback loop, a voltage-to-current converter. Therefore, the overall arrangement produces *torque* in response to *input voltage*. Although speed is at the mercy of the applied mechanical load, the speed regulation of permanent-magnet and shunt motors justifies the classification of these types as constant-speed machines. Of course, such a designation has the aspect of relativity—compared with series motors, the permanent-magnet and shunt motors are quite deserving of such classification. Also, they compare favorably to induction motors in this respect. However, compared with *synchronous motors*, such "constancy" would prove misleading for many applications.

The motor employed in such a torque-control system will actually display poorer speed regulation than it normally would without control. This is because any current-regulating process degrades the voltage regulation across the load through which the current is stabilized. Inasmuch as motor speed is proportional to armature voltage, it follows that the speed in such a system will tend to be an undisciplined performance parameter. However, this is generally not of detrimental consequence in many torque-control applications such as winders, unwinders, tension mechanisms, and others.

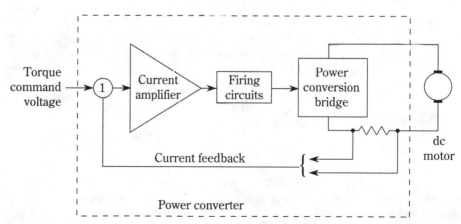

4-10 Torque control of a permanent-magnet or shunt dc motor. Randtronics, Inc.

A simplified schematic diagram of a power-conversion bridge suitable for the torque-control technique is shown in Fig. 4-11. The bridge is unidirectional in the sense that current always flows in the same direction through the motor. The waveforms pertaining to low- and high-torque conditions are illustrated in Fig. 4-12.

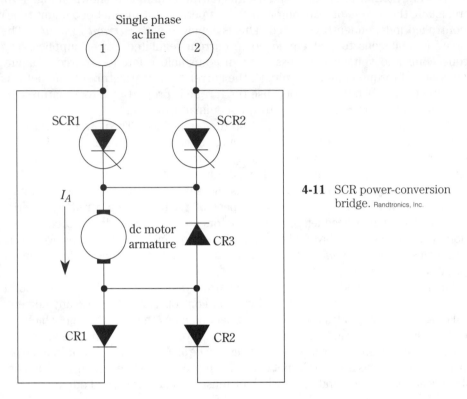

4-11 SCR power-conversion bridge. Randtronics, Inc.

Tachometer feedback system for speed control

A speed-control system cannot be more accurate than the method used for the actual sensing of motor speed. Although there is close correlation between impressed armature voltage and speed, and between counter EMF and speed, there are applications where more precise speed sensing is needed. This is readily attained by coupling a dc generator, or tachometer, to the motor shaft. Thus, a dc voltage representing speed is produced. This voltage can represent speed quite accurately and is not influenced by armature reaction, current, or temperature within the motor itself. Also, the polarity of the tachometer voltage changes with the direction of rotation. This is a fortuitous convenience in bidirectional systems. The tachometer signal is usually fed back to the input of the system where it is compared with a variable reference, or speed-command, voltage. The comparison produces an *error voltage* of the polarity required to extinguish itself by correcting the motor speed. And, as long as the speed-command voltage is held constant, the motor speed will be maintained constant despite variations in mechanical load or in other motor operating factors.

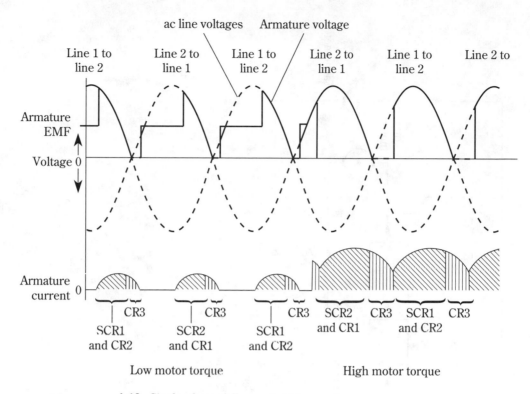

4-12   Single-phase, full-wave bridge waveforms. Randtronics, Inc.

Figure 4-13 is a block diagram of a speed-control system using tachometer feedback. There are other feedback loops as well. These are for the purposes of linearizing overall response, increasing bandwidth, and limiting motor current. The tachometer feedback loop embraces all the amplifier stages and therefore is instrumental in determining motor speed. The system shown is intended for use with either a permanent-magnet motor or a shunt motor. With a shunt motor, the high end of the speed range can be extended by reducing field current. However, the available torque will then be less for these higher speeds. In principle, a series motor could be used, but for many applications, the horsepower rating could easily be exceeded. If the motor size and the nature of the load are such that a safe temperature rise can be maintained in the motor, the tachometer speed-control method of Fig. 4-13 will work with any type of dc motor.

The configuration for the driver and output stages used in the tachometer speed-control system is shown in Fig. 4-14. Bipolar power is made available for the motor to provide for reversal. Dual power supplies are required, as are double outputs from the op-amp stages. The output stage functions in the linear mode and can accommodate a number of paralleled transistors providing that the current-sharing emitter resistances are used, and due consideration is given the matters of drive and heat removal.

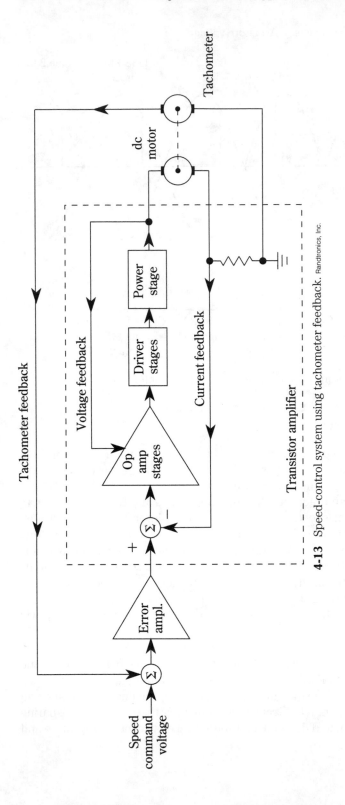

4-13 Speed-control system using tachometer feedback. Randtronics, Inc.

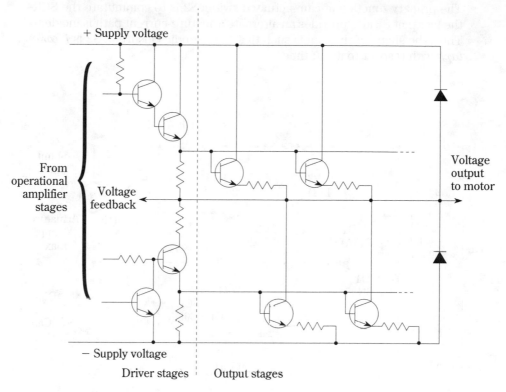

4-14 Driver and output stages for speed-control system of Fig. 4-13. Randtronics, Inc.

Synchronous speed control for shunt and permanent-magnet motors

By means of a unique method, the speed-versus-torque behavior of shunt and permanent-magnet motors can be stabilized to simulate the operation of the ac synchronous motor. At the same time, the basic features of dc machines, such as high starting torque, electrically adjustable speed, and operation from batteries, are retained. An obvious application area is magnetic-recorded equipment. There are also many timing techniques in photography and in industrial instrumentation where such a motor control is required.

Figure 4-15 depicts the circuit for achieving such speed regulation. The modification of motor behavior attained with its use is shown in Fig. 4-16. Note that full torque is available for starting. In the circuit of Fig. 4-15, the SCR is triggered at periodic intervals by a unijunction relaxation oscillator, Q1. A second feature of the circuit is the cam-actuated switch, S2, which is mechanically coupled to the motor shaft. The following conditions apply:

- The armature current path is either through the SCR or through switch S2.
- The closure of switch S2 not only established the armature current path, but commutates the SCR. That is, conduction in the SCR is stopped at the closure of S2.

- The primary function of cam-actuated switch S2 is to commutate the SCR—the fact that S2 also provides an alternate armature-current path is incidental. Thus, the shape of the cam is such that the periodic closure of S2 is *momentary* with respect to its off time.

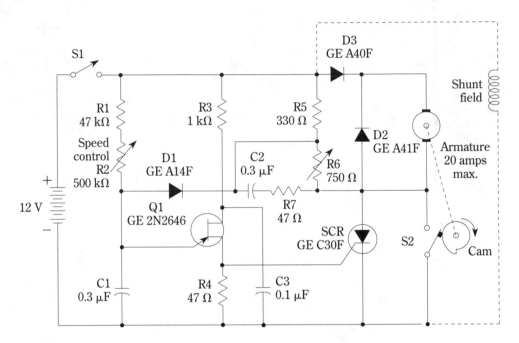

4-15 Synchronous control for permanent-magnet or shunt dc motors. General Electric Co.

If a load is placed on the operating motor, its natural tendency will be to slow down. However, such a slowdown results in longer conduction time for the SCR. In turn, this means higher average armature current, which accelerates the motor. The speed correction thus attained will not be greater than that permitted by the turn-on rate of the SCR. Because this turn-on rate is determined by the oscillator, the motor speed locks to the oscillator frequency. The opposite reactions occur if the motor load is relaxed. Referral to the timing diagram of Fig. 4-17 reveals that the oscillator frequency and the SCR turn-on rate remain constant over the full range of motor loading. The commutation of the SCR is, however, modulated in time. At first, the nature of these events might seem to be contradictory. How can the SCR be turned on synchronously with the motor rotation, but be turned off at apparently random intervals?

The answer is that loading of the motor causes a *momentary* slowing of its instantaneous speed, just enough to enable it to lag farther behind the oscillator while maintaining its average synchronous speed. In other words, the phase displacement between the two ac sources, the oscillator and the motor-actuated switch, widens.

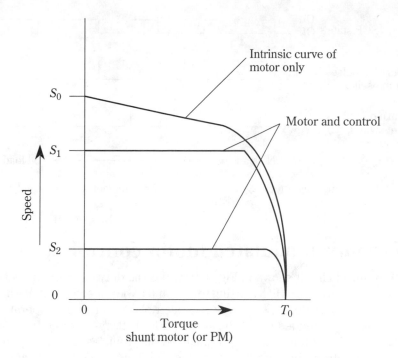

4-16 The effect of synchronous speed control on permanent-magnet and shunt dc motors. <small>General Electric Co.</small>

The consequence of this is a longer turn-on time for the SCR, enabling more torque-producing current to be delivered to the synchronously rotating motor. This sequence of events is directly analogous to the operation of an ordinary ac synchronous motor, in which an increased torque demand is satisfied by additional lag of the rotor with respect to the rotating field. The rotor, however, maintains an average speed that is synchronous with the rotating field.

In order to prevent hunting, false locking, and other types of erratic operation, the triggering of the SCR must be delayed when the motor speed is not close to synchronism. This is accomplished by the delaying network comprising R5, R6, R7, C2, and D1.

Diode D2 is the "free-wheeling" diode commonly found in switching-type power supplies. It provides a current path for the motor armature during off time, thereby smoothing the torque. It also prevents destructive arcing of the cam-actuated switch in this circuit (in other words, D2 provides a *useful* outlet for the magnetic energy stored in the motor).

The ambitious experimenter might consider the possibility of using a nonmechanical sensing technique. This can be accomplished with optoelectronic methods or by magnetic sensors (both of these suggested sensing approaches can be found in electronic ignition systems for automobiles). It would probably be wise to first obtain satisfactory operation with the mechanical scheme as described. Then, the modification to electronic sensing should prove to be straightforward.

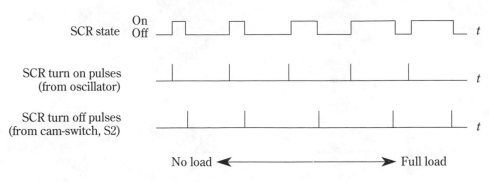

4-17 Timing diagram for synchronous control of a dc motor. General Electric Co.

Pulse-width modulated motor control

The motor-control circuit shown in Fig. 4-18 varies the duty cycle of current pulses delivered to a series motor. High starting torque and a wide speed range are available from this circuit. The design emphasis has been placed on overall operating efficiency because the intended use of the system is for battery-powered traction vehicles, such as golf carts, fork lifts, and small service vehicles. The motor receives "chopped" power from the 36-volt battery and can draw as high as 300 amperes of average current when working against a demanding load.

Although the schematic diagram shows a Darlington output stage, it is not possible to produce currents on the order of 300 amperes from single readily available transistors. Accordingly, the output stages of the Darlington amplifier consist of paralleled MP506 germanium power transistors. Each transistor in the parallel circuit should have its individual emitter resistor in order to equalize current sharing.

The "free-wheeling" diode, D10, is important in pulsed circuits where the load is inductive, such as in a motor. This diode enables motor current to continue during the off periods of the power-output stage. This current is due to the energy stored in the magnetic field of the motor during on time. This phenomenon has nothing to do with motor or generator action but stems only from the inductive nature of a motor as a load—a simple inductor or filter choke would behave in the same manner. The continued flow of motor current after the power-output stage has switched to its off state is not contradictory and does not defeat the switching process. The *average current* in the motor still remains a function of the ratio of on time to off time, but it has a smoother waveform than if there were no diode in the circuit. Moreover, if the diode were not used, the inductive-kickback voltage spikes occurring at the switching transitions would possibly damage the transistors.

The portion of the circuit labeled, *Pulse Modulator*, consists of a multivibrator, Q5 and Q6, driven by a unijunction relaxation oscillator, Q4. This circuit generates the controllable duty-cycle pulse train which, after being amplified by stages Q7 and Q8, drives the Darlington power-output stage. Except for a current-limiting provision, there is no feedback in this control system. The duty cycle, and therefore the average current delivered to the motor, is manually controlled by potentiometer R17.

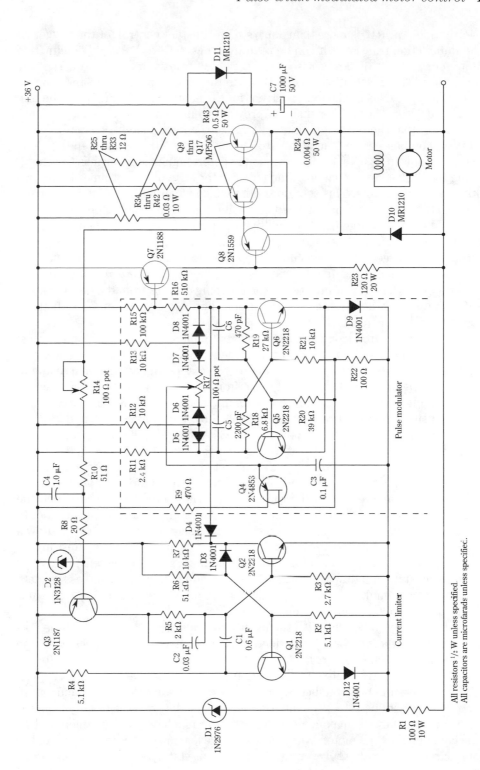

4-18 Pulse-width modulated motor-control circuit. Motorola Semiconductor Products, Inc.

All resistors 1/2 W unless specified.
All capacitors are microfarads unless specified.

Potentiometer R17 is more than a mere time-constant control for the triggering of the unijunction transistor. It can be seen from Fig. 4-18 that R17, in conjunction with its four steering diodes, is associated with *both* transistors (Q5 and Q6) of the multivibrator. This association is such that the turned-off transistor of the multivibrator always provides the charging current for the emitter capacitor, C3, of the unijunction oscillator. The result is that R17 provides adjustment of the on and off times in such a way that the pulse-repetition rate remains approximately constant.

In one complete cycle of multivibrator action first one transistor then the alternate transistor is in its off state. The frequency-determining mechanism is the charging time for C3. The total charging time for C3 is the *sum* of the off times in the two multivibrator transistors. This sum does not change when R17 is adjusted. For example, if R17 is at one extreme of its adjustment range, capacitor C3 could take a relatively long time to charge from the multivibrator transistor that happened to be in its off state. Once charged to the triggering potential of the unijunction transistor, C3 is abruptly discharged and the resulting trigger pulse *reverses* the state of the multivibrator. Now C3 charges up relatively fast because the *minimum* resistance arm of potentiometer R17 is now involved in the charging process from the other transistors. Therefore, the duration of a cycle is not altered—only the on and off times change with adjustment of R17.

The remaining circuitry is that of the current limiter. Current limiting is necessary in the intended applications because of the willingness of a series motor to exert torque regardless of the opposition. In order to extend the battery charge and to protect both the motor and the output transistors, a means is provided for automatically extending off time when a preset maximum motor current is exceeded.

The sensing of motor current is accomplished by pnp transistor Q3 and its associated tunnel diode, D2. Ordinarily, transistor Q3 is deprived of base-emitter forward bias by the shunting action of the tunnel diode. However, the tunnel diode is triggered to a higher operating voltage if it is subjected to current exceeding its "peak" value. This action is very fast and precise. It occurs as a manifestation of the negative-resistance characteristic of the tunnel diode. The motor current at which the tunnel diode triggers is adjustable by control R14. When the tunnel diode switches to its higher-voltage state, transistor Q3 receives sufficient base-emitter bias to draw a large collector current. This event triggers a monostable multivibrator comprising Q1 and Q2. Ordinarily (when the motor is *not* demanding excessive current) transistor Q1 is in its on state and transistor Q2 is off. The conduction of Q3 reverses the state of the monostable multivibrator—but only for a time determined by the RC values associated with the monostable circuit.

The temporary state reversal of the monostable multivibrator interrupts the normal operation of the previously described pulse-modulator circuit. Specifically, the pulse-modulator circuit comprising Q4, Q5, and Q6 is overridden, and transistor Q7 remains in its conductive state for the duration of the monostable multivibrator cycle.

When transistor Q7 is conducting, the power-output stage delivers *no* current to the motor. This action protects the motor, the output stage, and the battery from excessive current. Only the maximum current as determined by the adjustment of R14 can be consumed by the motor. When the excessive load on the motor is relaxed, the sensing circuit (Q3 and D2) reverts to its quiescent state, as does the monostable multivibrator, Q1 and Q2. There is then no interference with the normal operation of the pulse-modulator circuits involving Q4, Q5, and Q6. Therefore, motor control is again completely governed by the setting of the "throttle," R17.

The design philosophy of this control system is worthy of study in view of the growing emphasis on electric vehicles. The use of germanium power transistors is interesting. These devices have certainly not been relegated to obsolescence by silicon devices. Germanium transistors are inexpensive, and because of their low collector-emitter saturation voltage, they operate more efficiently in some high-power pulse applications that do silicon transistors. Additionally, they do not have the commutation problems inherent in dc applications of SCRs.

Motor speed control with a phase-locked loop

The speed-control scheme shown in Fig. 4-19 is similar to servo systems, feedback circuits, and regulated power supplies in one respect. All of these are "automatic mulling" applications where the error between a sampled portion of the output signal and a *reference signal* is reduced to zero. In order for the error signal to become zero, the output signal is forced to return to the level from which it attempted to deviate. Thus, in a simple regulated power supply, any tendency of the output voltage to fall produces an error signal because the sampled portion of the output voltage would then differ from the reference voltage. Such an error signal would then increase the conductivity of the series-regulator transistor, which would in turn *increase* the output voltage until the error signal was restored to zero (a tendency for the output voltage to rise would invoke the opposite reaction). A regulated dc power supply is shown in Fig. 4-20. Its operation provides a good *analogy* for understanding the phase-locked loop.

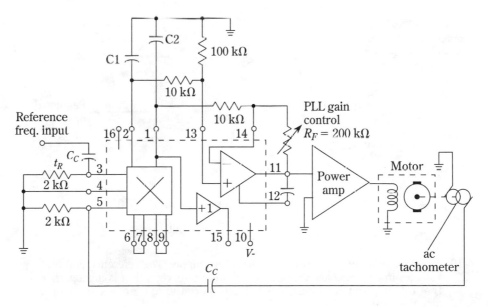

4-19 Speed-control by means of a phase-locked loop. Exar Integrated Systems, Inc.

The salient features of the regulated power supply, analog servo systems, and many other self-corrective circuits are:

- Negative feedback is used—a sample of the output is returned to the input.
- At the input, the feedback signal is *compared* with a reference voltage, and the difference between the two generates an error signal.
- The error signal is amplified and deployed to change the output signal in the direction tending to *extinguish the error*.

The phase-locked loop is also a self-correcting system. It differs in the following ways from the more familiar analog, or "linear," system just described:

- The reference signal is an *ac* signal.
- The error signal is an ac voltage or a *digital pulse train*. (Departure from the reference frequency is the "error.")
- Regulation involves the stabilization of an output frequency, which can be made to represent the *speed* of a motor.
- The basic circuit action is to cause the *frequency* of the sampled output to become identical to that of the reference frequency.

The basic phase-locked loop is illustrated in Fig. 4-21A. This arrangement is often found in communications systems. Thus, if an FM signal is applied to the phase comparator, the filtered error signal constitutes the audio modulation. The phase-

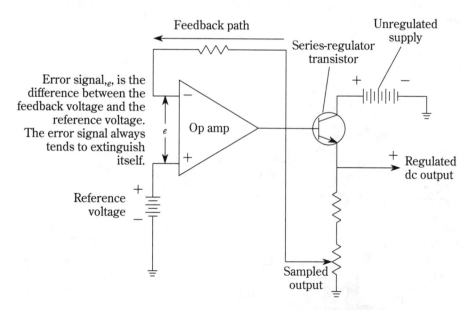

4-20   Error-signal nulling action in a regulated dc supply. The reference and error voltages in this case are *dc*, but the nulling or servo action is analogous to that taking place in a phase-locked loop where the reference and the error signal are *ac*.

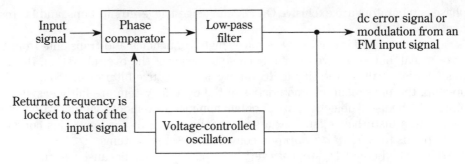

A. Basic phase-locked loop.

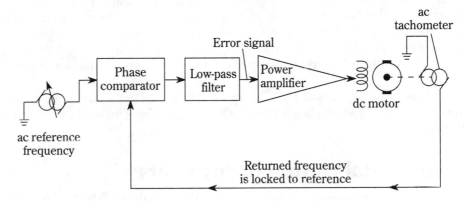

B. Phase-locked loop for speed control of a dc shunt motor.

4-21 Simplified block diagram of phase-locked loop systems.

locked loop in such an application serves as a frequency discriminator. The circuit action is such that the output frequency of the voltage-controlled oscillator always seeks the instantaneous frequency of the FM signal. The phase-locked loop therefore displays selectivity and can be used in place of conventional tuned circuits. The error signal is stripped of its high-frequency residue by the low-pass filter. Because of its position in the loop, the voltage-controlled oscillator is caused by the error signal to make its generated frequency *identical* with that of the incoming signal. It can also be said that the action is such that the *error signal tends to extinguish itself,* as it does with analog servo systems. Now, consider the adaptation of this basic system to the control of *motor speed*. Figure 4-21B is functionally the same as the integrated circuit system in Fig. 4-19.

Figure 4-21B shows that the electronic voltage-controlled oscillator is replaced by a dc shunt motor coupled to an ac tachometer. The tachometer can be a small ac generator, or it can be an optical encoder formed from a slotted wheel in conjunction with a light source and a photoelectric detector. (This could be an LED and a phototransistor.) For simplicity, the power supply for the motor armature is not shown.

A number of circuits can provide the function of phase comparator. In Fig. 4-19, a four-quadrant multiplier within the IC is used as the phase comparator. Other phase comparators are exclusive OR circuits, edge-triggered flip-flops, and LC resonant circuits.

The low-pass filter not only removes high frequencies and transients from the error signal, but also governs the dynamic behavior of the overall system. Because this filter is usually of the simple RC variety or is an active filter designed around an op-amp, the phase-gain characteristics of the overall system are fairly easy to manipulate. The usual objectives are to obtain minimal over-shoot together with fast response to a disturbance, such as a suddenly increased motor load. This condition corresponds to *critical damping* in conventional servo systems.

This speed-control system has very good potentialities, because extremely close speed regulation is possible with its use. It makes feasible such techniques as coordinating motor speed with digital clocks (which is needed in certain computer peripherals) or synchronizing the speed of several motors in a conveyer system. And once the basic operation of this system is grasped, its use can be extended to other types of motors. Alternating-current motors could conceivably be used if a voltage-controlled oscillator is inserted between the low-pass filter and the power amplifier in Fig. 4-21B.

Although the circuits of Fig. 4-19 and 4-21B control the field of a shunt motor, armature current can also be controlled. Of course, a more powerful amplifier would be needed.

Incremental control of dc motors

The incremental control of dc motors involves digital feedback. The feedback signals contain quantitative information concerning the angular position of the motor. This type of control is probably the most sophisticated and advanced of all electronic techniques for controlling motor operation. Because of the precise behavior that can be imparted to the shaft, the incremental control of motors is intimately associated with the automation of industrial processes and, in particular, with the numerical control of machine tools from computer programs.

A simplified block diagram of an incremental motor control system is shown in Fig. 4-22. The forward/reverse counters function in a way analogous to the *comparator* in a linear servo system or in a regulated power supply—an error signal is generated when there is a difference between feedback information and the information provided from a reference. In this system, as in many servo and regulator systems, control is achieved by varying the reference information from an external source. Thus, a command imparted to the *frequency generators* provides a new "reference," causing the incremental control system to seek a new equilibrium. In so doing, the motor shaft turns through an accumulated angular distance exactly as commanded. In another mode of operation, the motor can be commanded to rotate at a precise speed. Furthermore, the acceleration and deceleration of the motor is controllable, as is "jogging," starting, stopping, and reversing. Such control versatility stems from the *positional information* carried by the feedback path.

The basic operating principle of this control system involves the subtraction of feedback pulses from accumulated command pulses in the forward/reverse counters (otherwise known as an "error register"). This subtraction produces the position er-

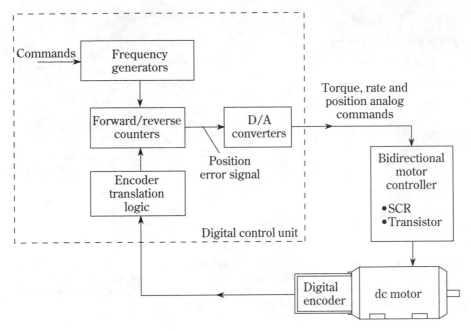

4-22 Simplified block diagram of an incremental control system. Randtronics, Inc.

ror signal. As in any servo or regulator system, the error signal invokes a sequence of events causing the signal to be extinguished. In this system, the error signal is converted to a voltage level by the digital/analog converter. Then it is boosted to a power level suitable for actuation of the motor. Motor actuation always has as its "objective" the *reduction to zero* of the positional error signal appearing at the output of the forward/reverse counters.

If the command consists of a pulse train with a constant pulse-repetition rate, the motor speed will be constant. Long-term accuracies on the order of several parts per million are readily achieved in this manner. On the other hand, a predetermined number of command pulses can be used to position-synchronize a motor-driven mechanism to an external source.

The Kramer speed-control system

The three-phase wound-rotor induction motor exhibits the advantage over other induction motors in that its speed can be adjusted over a considerable range. On the other hand, it suffers the disadvantage in that its speed regulation is poor. That is, a variation in the loading tends to produce a large change in speed. Additionally, the speed-control rheostats in the rotor circuit sometimes waste considerable power. There is a relatively simple way to obtain up to a four-to-one speed range with good speed regulation, and without high-power dissipation. The arrangement shown in Fig. 4-23 is known as the Kramer drive. A dc shunt-field motor shares the same shaft as the wound-rotor induction motor. Electronic circuitry is limited to rectification in the basic setup.

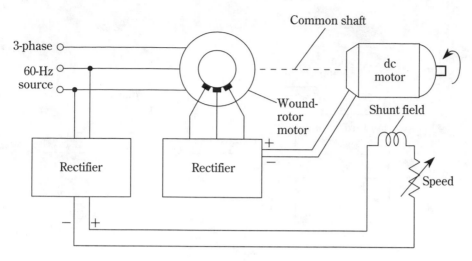

4-23 The speed-control system. The combination of ac and dc motors coupled to a common shaft provides an efficient adjustable-speed range with good speed regulation. Speed changes from varying the load on the output shaft are much smaller than would result from either motor alone.

In this unique scheme, the ac slip-power from the rotor of the induction machine is rectified and delivered to the armature of the dc motor. Speed control of this combination is obtained by adjusting the field current to the dc motor. Inasmuch as field current in a shunt dc motor is, at most, a small fraction of its armature current, there is little opportunity for power wastage. Speed regulation with respect to load change is very good because of a feedback mechanism; suppose that the load is increased, tending to slow the rotation of the shaft. The lower speed increases the slip power of the induction motor, which causes greater current to be delivered to the armature of the dc machine, thereby forcing it to speed up. The net effect is that the shaft speed is considerably stabilized against load changes. Note that both machines contribute torque to the common shaft.

There are other ways of viewing this duo. It bears some resemblance to the concatenation of induction motors described elsewhere. This stems from the scheme of having two motors coupled to a common output shaft. The wound-rotor induction motor can be thought of as a variable transformer with the rotor serving as the secondary. Also, the dc shunt motor performs somewhat as a magnetic amplifier in that the control of low power in its field circuit serves to control high power in its armature circuit. In any event, it is noteworthy that there are no rheostats in heavy-current portions of the circuit.

The maximum speed of this technique is that of the induction motor. A commonly attained speed range for 60 Hz is 450 to 1700 RPM. The minimum speed is a consequence of the relationship between voltage provided by the rotor of the induction motor and that needed by the armature of the dc motor. This relationship tends to be contradictory to extending the speed-control range much beyond the alluded four to one.

Dynamic braking circuit for dc motors

When a motor drives a load that is essentially inertial in nature, a long coasting period might ensue when the motor is turned off. This can be undesirable for certain applications. Slowdown by mechanical braking often proves both awkward and costly. *Dynamic braking* is a much used technique in which the motor is operated as a generator after being disconnected from its source of electrical power. When a machine is functioning as a motor, it develops a counter EMF; when the same machine operates as a generator, it develops a *counter torque*. The more electrical power extracted from such a generator, the greater the countertorque produced in its armature. Therefore, if the armature terminals of a dc motor were immediately shorted following turn-off, a very rapid deceleration of the rotating shaft would be anticipated (providing the field remained energized).

In order to spare the armature from undue mechanical stress, as well as electrical stresses on the commutator and brushes, the armature short circuit is not often used. Rather, a low resistance connected across the armature suffices to bring the motor to a quick enough halt. The resistance power rating must be able to absorb the sudden conversion of kinetic energy to electrical energy and dispose of the resultant heat.

Figure 4-24 shows a novel electronic circuit that *automatically* provides dynamic braking when a dc motor is shut off. This circuit is intended for use with permanent-magnet motors but is also adaptable to shunt-wound and some series-wound motors. When switch S1 is in its RUN position, dc power is fed to the motor armature

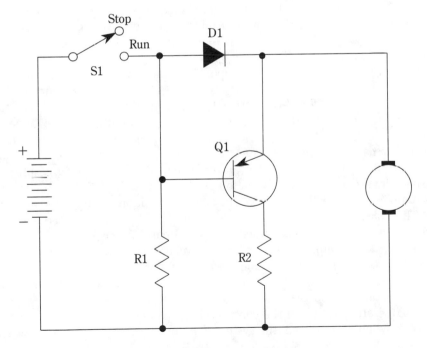

4-24 Automatic braking circuit for dc motors.

through rectifier D1. The transistor circuit is inactive because the emitter base is reverse biased. When switch S1 is placed in its stop position, the still rotating armature generates a voltage of the same polarity as the power source, insofar as the emitter and collector of the transistor are concerned. The base is, however, now *isolated* from the emitter because of the unidirectional characteristic of diode D1. This being the case, the base becomes heavily forward biased through resistance *R1*. Transistor Q1 is driven into saturation and behaves as a closed switch, connecting resistance *R2* across the armature. This causes the motor to relinquish its mechanical motion in order to "pay" for the heat energy developed in resistance *R2*.

If a shunt motor is used, the field must remain energized during the slowdown period. Otherwise, only a negligible reduction of coasting time will occur. A four-terminal series motor can be dynamically braked by this circuit if its field is energized from an appropriate source after the armature is disconnected from the line.

dc motor speed/position control system

Use of the stepper motor is not the only way to devise a speed/position control system. A PM dc motor can also be made to obey speed commands and to stop precisely at a predetermined position. This enables greater torque and power than is readily attainable from stepping motors. At the same time, very high resolution is possible. What is needed is a shaft-mounted optical encoder and some special electronics. The latter requirement can be conveniently satisfied with several "building-block" dedicated ICs, and a microprocessor or microcomputer. Overall, a servopositioning system results that benefits from closed-loop sensing and response (stepper motor positioners are conventionally open-loop systems).

The optical encoder is a special type of tachometer that generates more information than merely pulses proportional to speed. The nature of the encoder is illustrated in the sketch of Fig. 4-25. Notice that there are two slotted disks, a stationary one, and one rotated by the motor shaft. This is essentially a light-chopper technique. The slots in the two disks overlap so that two quasi sine-wave signals are generated that have a 90-degree phase relationship. The frequency of these signals represents the speed of the motor, and their plus or minus 90-degree phase relationship conveys directional information. Additionally, a third signal of one pulse per revolution is generated in order that the absolute position can be found at initialization. The quasi sine-wave signals are designated as the FTA and FTB signals. The pulse-per-revolution signal is termed the FTF signal.

The block diagram of the speed/position control system is shown in Fig. 4-26. The salient aspect of such a scheme is what could otherwise be a very complex array of electronic circuitries is made quite simple and straightforward with the use of the three dedicated ICs.

The operating modes of the control system can be gleaned from inspection of Fig. 4-27. When first started, the system is in its *speed-control* mode; however, at

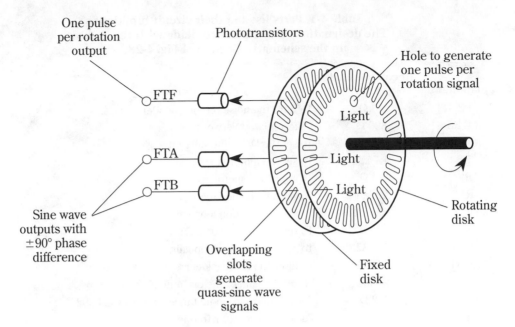

One pulse per rotation output

Phototransistors

Hole to generate one pulse per rotation signal

FTF

Light

FTA

Light

FTB

Light

Sine wave outputs with ±90° phase difference

Overlapping slots generate quasi-sine wave signals

Fixed disk

Rotating disk

4-25 Optical encoder for enabling a dc motor to operate as a servo positioner. Two 90-degree displaced signals are generated together with a one-pulse-per-revolution signal. SGS Thomson Microelectronics, Inc.

startup, the tachometer signal is either absent at zero-time or relatively small during acceleration. During this interval, the system is effectively operating open loop, enabling a high starting current. This is desirable because it promotes rapid speed buildup. Once the motor attains its preset level of speed, closed-loop speed regulation sets in and the speed of the motor is held constant.

Later, when the microprocessor commands deceleration in preparation for stopping at a predetermined position, the motor is progressively braked, reducing its speed. Finally, the system is commanded to switch to its *position-feedback* mode. The motor then stops abruptly at the desired position and is held there by an electronic detent.

The schematic diagram of the dc motor speed/position control system is shown in Fig. 4-28. Aside from the addition of a few passive components, it is almost as simple as the block diagram of Fig. 4-26. The apparent complexity that one sees at first glance stems from the depiction of the internal circuit functions of the three-dedicated ICs. It turns out that this draftsman technique is useful in analyzing the operating principles of the overall system. In particular, the circuit-point designations shown inside of the L290 coincide with the waveforms designations of Fig. 4-29.

Further assistance to circuit analysis is provided by the unique parts list of Table 4-1. Each part is accompanied by a brief statement indicating its purpose in the schematic diagram.

**Table 4-1. Parts list and their circuit functions.
The designations of the parts coincide with those shown
in the schematic diagram of Fig. 4-28.**

Component	Recommended value	Purpose
R1, R2, R3	1 kΩ	To filter the noise on the encoder signals.
R4, R5	820 Ω	Differentiator Network
R6, R7	4.7 kΩ	To set the D/A input current.
R8	4.7 kΩ	To set the motor speed.
R9	5 kΩ	To adjust the motor speed.
R11	22 kΩ	To set the position loop gain.
R12	100 kΩ	To set the position loop gain.
R13	120 kΩ	To set the speed loop gain.
R14	15 kΩ	To set the position loop gain.
R15, R16	510 Ω	To filter the feedback current.
R17	22 kΩ	To set the gain of the err. amplifier.
R18, R19	0.2 Ω	To set the transconductance value of the L292.
R20	15 kΩ	To set the oscillator frequency.
R21	33 Ω	Compensation Network
C1, C2, C3	100 pF	To filter the noise on the encoder signals.
C4, C6	15 nF	Differentiator Network
C5	2.2 µF	By-pass capacitor
C7	0.1 µF	Low-pass filter for the D/A input current.
C8	0.22 µ	—Low-pass filter for the tacho signal. —To determine the dominant pole of the speed loop.
C10, C11	0.1 µF	Supply By-pass capacitor
C12	47 nF	To filter the feedback current
C13	47 nF	To set the gain of the error amplifier C13 - C17 = L_M/R_M
C15	0.1 µF	Supply By-pass capacitor
C16	470 µF	Supply By-pass capacitor
C17	1.5 nF	To set the oscillator frequency and the dead time of the output transistors.
C18	1 nF	Compensation Network
D1, D2,	1 A	Recirculation Diodes
D3, D4	Fast Diodes	

A word is in order regarding the switching frequency of the L292 switchmode motor driver. This is governed by timing components R20 and C17. The best frequency will usually be found to be in the 16-kHz to 23-kHz range, where optimum balance between low noise and high switching efficiency is likely to be realized. However, capacitor C17 also sets the delay time that protects against simultaneous conduction of the output transistors. The suggested value of C_{17} is 1.5 nanofarads. This

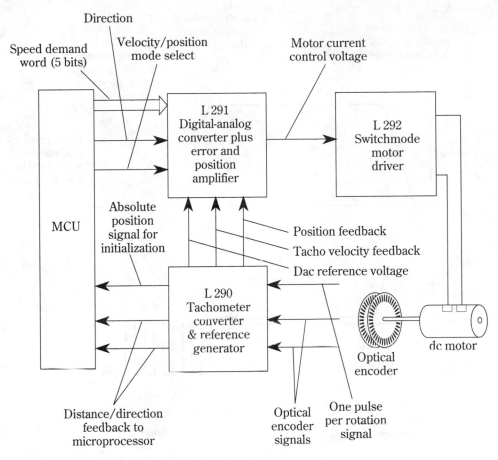

4-26 Speed/position control system for a dc motor. The dedicated ICs used in this scheme enable operation from an ordinary dc PM motor to simulate that of a stepping motor, but with possible advantages of torque and power. Very high positioning precision is another plus feature. SGS Thomson Microelectronics, Inc.

results in a protective switching delay of about 2.5 microseconds—more than adequate. A further stipulation in selection of the RC timing values is that R_{20} must be at least 8.2 kΩ. These restrictions allow sufficient flexibility in attaining optimum operation. The actual switching frequency will be approximately:

$$f_{\text{osc.}} = \frac{1}{(2)(R_{20})(C_{17})}$$

Under most circumstances it is better to purchase, rather than construct, the optical encoder. A suitable one is the Sensor Technology STRE 1600, which has 200 slots. Similar types are available from Sharp Electronics Corporation, Mahway, New Jersey.

Although the nominal 2-ampere, 36-volt output of the L292 driver is a respectable power level for an IC, it is understandable that greater drive for larger mo-

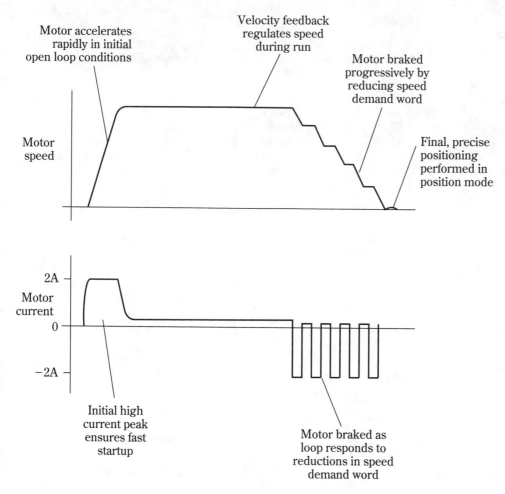

4-27 The operating modes of the dc motor speed/position control system. *Tachometer-feedback* regulates the speed during a run. It also provides for deceleration upon approach of the target position. Finally, *position-feedback* enables the motor to stop at the programmed position. SGS-Thomson Microelectronics, Inc.

tors might be needed. The best approach is to insert the required high-power stage between the L292 and the larger motor. In that way, the basic operating mode of the described system will not be altered. Because of the relatively low switching frequency, bipolar transistors, (including Darlingtons), power MOSFETS, and IGBTs all merit consideration as output devices. Low-inertia motors such as the pancake or printed-circuit types are best suited in position-servo systems.

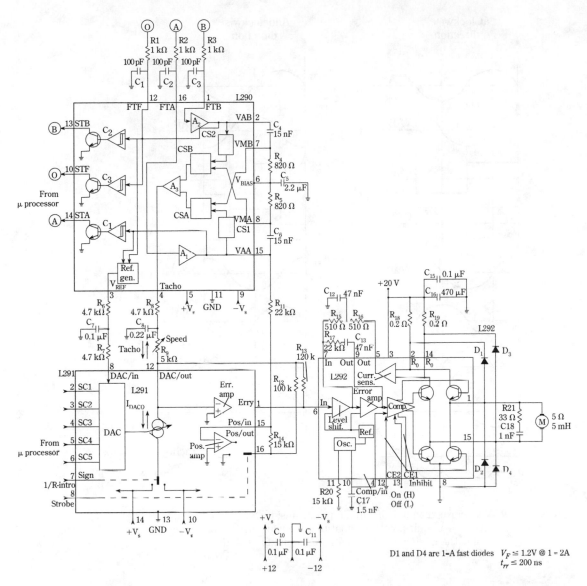

4-28   Schematic diagram of the dc motor speed/position control system. The functional components of the three building-block ICs are shown to facilitate understanding of circuit operation. Note that the L290 IC has designated circuit-points that coincide with those listed in the waveform diagram of Fig. 4-29.

SGS-Thomson Microelectronics, Inc.

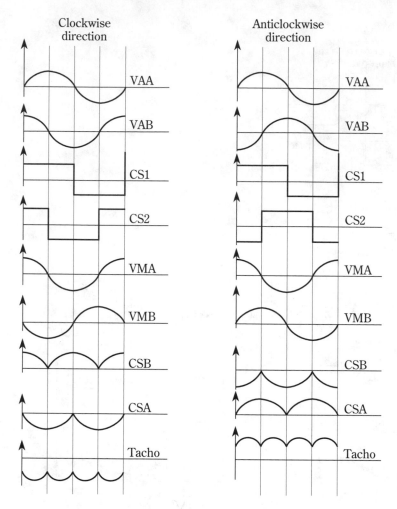

4-29 Waveform diagram of circuit points inside the L290 IC. Note that the circuit-point designations coincide with those shown inside the L290 in the schematic diagram of Fig. 4-28.
SGS Thomson Microelectronics, Inc.

5
Electronic control of noncommutator machines

THE NONCOMMUTATOR ELECTRIC MACHINE HAS ALWAYS BEEN FAVORABLY ASSESSED FOR its basic simplicity, with its attendant ease of manufacture, and for its exceptional reliability and relative freedom from radio frequency and electromagnetic interference. Some of these machines do have sliding contacts, but they are in the form of slip rings rather than commutators. Moreover, it is often true that the currents handled by the slip rings are much lower than those necessarily associated with commutators. Thus, in the automobile alternator, slip rings are used for conducting the field current to the rotor. This current is a small fraction of the charging currents that these alternators must handle. On the other hand, the older commutator-type dc generator used in automobiles had to pass the large charging currents through its commutator. As might be expected, the maintenance problem was far from trivial.

A shortcoming of noncommutator motors, however, was their inability to readily vary their speed over a wide range. Now, with solid-state electronics, this disadvantage is no longer necessary. The new control techniques give old-style noncommutator motors a performance flexibility never dreamed feasible by their original designers.

The following control circuits are interesting because they overcome the performance limitations long held to be inherent in ac machines, particularly induction motors. Also, you can sense the keen competition surrounding the selection of motor types. With the new control techniques, it no longer suffices to consult a motor text, or even motor specifications. To a large extent, you can now electronically "tailor" machine characteristics. Therefore, decisions must be influenced more by other factors, such as cost, reliability, electrical and noise considerations, etc.

Triac speed-control circuit for induction motors

The triac speed-control circuit for induction motors shown in Fig. 5-1 is similar to that shown in Fig. 4-3, which is intended for use with universal motors. The circuit in Fig. 5-1, however, incorporates a *single*-time-constant circuit to delay the phase of the gate trigger. This simpler approach is permissible because induction motors

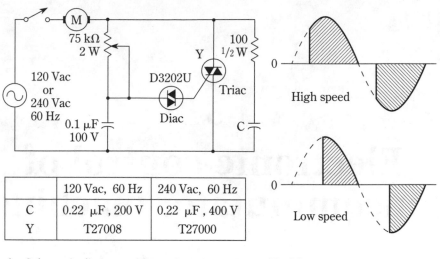

	120 Vac, 60 Hz	240 Vac, 60 Hz
C	0.22 μF , 200 V	0.22 μF , 400 V
Y	T27008	T27000

A. Schematic diagram with components listed for two different line voltages.

B. M

5-1 Triac speed—Control circuit for induction motors. RCA.

generally cannot be slowed down enough to get into the troublesome hysteresis area for which the double-time-constant gate circuit is prescribed as a remedy. This speed-control circuit works best for the permanent split-capacitor type of induction motor. The shaded-pole induction motor is also amenable to this control technique. With any type of induction motor, this speed-control technique is most effective when the load is a fan or blower. (A small change in speed produces a relatively large change in *air velocity.*) Another favorable aspect of such loads is their low starting-torque requirements.

Resistance-start and capacitor-start induction motors can be triac controlled under certain conditions. Generally it will be necessary to *limit* the speed-control range; the speed should not be reduced to the point where the centrifugal switch reconnects the starting winding or starting capacitor. All things considered, the greatest range of speed control will be obtained with the permanent split-capacitor motor. This type of induction motor is not encumbered with a centrifugal switch. Moreover, it operates well in the high-slip region. A three to one speed-control range is possible with fan loads.

This circuit is greatly superior to the single-SCR, phase-controlled thyristor circuit for use with induction motors. The SCR works well with *universal motors*, but the dc component developed by half-wave rectification is detrimental to the operation of induction motors.

The RC "snubbing network" connected across the triac does not generally appear in the circuit when the load is resistive, which is the case with lamps or heaters. Because a motor load is inductive, triac turn-off will occur at zero current, but the voltage across the triac will not be zero at that time. A voltage step is thus developed across the triac that can cause retriggering, despite the lack of a gate signal. This can happen even if the voltage-blocking capability of the triac exceeds peak ac voltage by a comfortable margin. The culprit is not necessarily the magnitude of this voltage step, or "spike," but rather its rate of change. Triacs specified with a high dv/dt across the main terminals will, other things being equal, tend to reduce the likelihood of such malperformance.

Motor-reversing technique for garage door openers

Figure 5-2 illustrates a simple, but reliable, technique for reversing the motor that actuates electronic garage door openers. The motor used is a permanent-capacitor induction type. Such a single-phase machine is admirably suited to this application. Rotation occurs in one direction when the first triac is turned on, and in the reverse direction when the alternate triac is turned on. Because of the full-wave conduction characteristic of triacs, it is as if a switch were closed when either triac is triggered into its conductive state. The physical nature of the system precludes the possibility of both triacs conducting simultaneously.

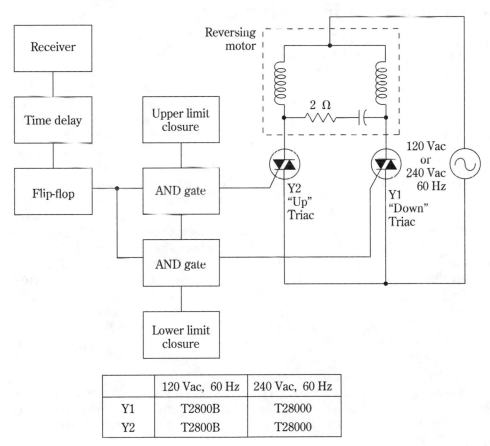

	120 Vac, 60 Hz	240 Vac, 60 Hz
Y1	T2800B	T28000
Y2	T2800B	T28000

5-2 Motor-reversing technique for garage door systems. RCA.

Various other motors have been used for garage door openers. These include capacitor-start and resistance-start split-phase types, as well as repulsion and repulsion-induction motors. However, the permanent-capacitor motor is best adapted to the simple triac circuit of Fig. 5-2. The capacitor size can be somewhat larger than normal so that starting torque will be increased. There are also split-field *universal* motors available that will operate with this scheme (it is not always considered good practice to use this type of motor in a belt-driven system, although most ¼-hp or ⅓-hp units will not race to destruction in the event that the belt is lost).

Garage door openers have features not evident from the motor-control circuit in Fig. 5-2. In order to avoid actuation from other transmitters, the UHF or VHF signal is usually tone modulated. Selective resonant circuits associated with the receiver then respond only to the intended transmitter. Mechanical devices disengage or reverse the drive mechanism in the event that an object impedes the motion of the door. This prevents injury to children or animals. A light usually remains on for about two minutes after the door is opened or closed. These features do not make any imposition on the motor, however.

Electronic switch for the capacitor-start induction motor

A clearly definable goal of electronic evolution has been the replacement of *mechanical* devices with solid-state devices. The basic idea, of course, is to improve reliability. As often as not, enhanced performance also attends the changeover from mechanical to electronic operation. This is because more precise timing can generally be attained, and operation occurs in more favorable surroundings. With regard to ac motors, it has often appeared somewhat incongruous to use a centrifugal switch in the capacitor-start induction motor; part of the rationale for selecting the induction motor is to *eliminate* switching contacts. When you use mechanical switching, you will generally be subjected to more frequent maintenance.

The scheme illustrated in Fig. 5-3 dispenses with the centrifugal switch, substituting a triac instead. The gate circuit of the triac is coupled to the ac line by means of a current transformer. The primary of this transformer consists of one or more turns of heavy conductor so that motor operation is not appreciably affected by inserting the transformer primary in series with the line. The secondary winding has a selected number of turns so that the triac is triggered by the inrush of current through the running winding of the motor. When the triac is turned on, the starting winding is energized. As the motor approaches running speed, the current through the primary of the current transformer is no longer sufficient to cause the triac to be triggered through the secondary winding. As a result, the triac turns off and the power is no longer supplied to the starting winding. Capacitor C1 and resistor R1 form a snubbing network to prevent inadvertent triggering of the triac from the inductive kickback of the starting winding.

This simple electronic switch can often be placed inside the motor itself. Not only does it effectively substitute for the centrifugal switch used on most capacitor-start induction motors, but it similarly replaces the *current* relay used for the same

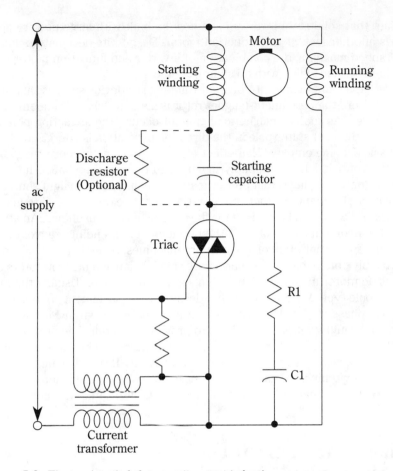

5-3 Electronic switch for capacitor-start induction motor. General Electric Co.

purpose on many of these motors. Aside from the obvious advantages of eliminating mechanical switches, the freedom from sparking can be a worthwhile feature when the motor is operated in a volatile atmosphere.

Brushless dc motors

The desire to devise electrical machines without sliding contacts has resulted in experimental and commercial "brushless dc motors." Most of these motors have limited torque capability—they are usually specified for use with fans or various instrumental applications. However, work has also been done along this line with regard to large motors and generators. The objective, in any case, is to eliminate the maintenance factor inherent with commutators and brushes. The controllability of dc machines inspires their selection, whereas the reliability of the commutator brush assembly is an adverse factor. Thus, it has been natural to desire the merger of the flexible operating characteristics of the dc motor with the simplicity of the squirrel-cage ac motor. For example, the fact that solid-state inverters can now be built with large power capacities demands another look at the induction motor. Despite its designation as a near-constant speed machine, the combination of inverter and induction motor now yields a variable-speed machine with no sliding contacts.

Perhaps more in line with present thinking is a device such as that illustrated in Fig. 5-4. Here the permanent-magnet rotor is made to "chase" sequentially switched stator poles. As soon as magnetic alignment occurs, the attracting pole is de-energized and the next stator pole in the direction of rotation is energized. A variation of this basic scheme employs Hall-effect elements to directly sense the position of either the rotor itself, or a suitably magnetized disk on the shaft. The advantage of such magnetic sensing is that power need not be dissipated in light sources. The toy-like aspect of such no-contact commutation might be deceptive. Circuits using optic sensors and SCRs can be devised for the control of giant machines. An attractive feature of noncommutator motors is the possibility of extending control to include positioning, synchronization of two or more machines, reversing and braking.

Readily obtainable on the market are small dc fan and blower motors comprising a small ac motor in conjunction with a solid-state inverter. Usually the motor is the shaded-pole type. More sophisticated designs use a permanent split-capacitor motor or a two-phase motor. The general idea involved is shown in Fig. 5-5. Sometimes, leads are brought out of the package to provide for optional ac or dc operation. The result is a "universal motor" without the commutator and brushes. A compelling feature of these inverter motors is their relatively low RFI. It is true that the inverter generates switching spikes. However these spikes are more amenable to manipulation and filtering than are the sporadic transients associated with commutator sparking and arcing.

Triac control circuit for three-phase induction motors

The arrangement illustrated in Fig. 5-6 is a solid-state relay circuit for three-phase inductive loads, such as induction motors. On and off control of the motor is affected by a low-level logic signal applied to the input of the system. Optoelectronic devices are used in the input circuits for electrical isolation between the control logic and the three-phase power circuitry. It can be appreciated that turning a large polyphase motor on and off is not a trivial problem. Mechanical or electromechanical devices are bulky, expensive, and high maintenance prone. Moreover, the arcing associated with physical contacts generates RF interference and constitutes a hazard in some operating environments.

The CA3059 IC modules are actually zero-voltage switches—that is, they have the property of providing triac gate pulses only when the voltage wave of one of the three phases crosses its zero axis. Such thyristor triggering results in minimal RFI. However, these IC modules are not employed in their usual fashion in this circuit. It is not necessary, or desirable, to attempt zero-voltage switching when the load has appreciable inductance. The inductance, itself, prevents abrupt current transitions on "make" (the slowed down current buildup serves as an RFI preventative). The

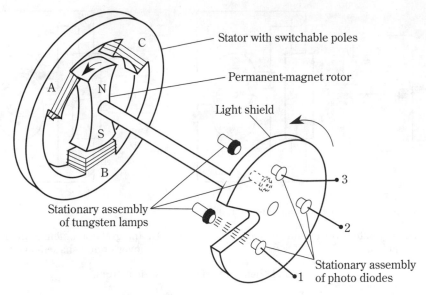

A. Mechanical arrangement.

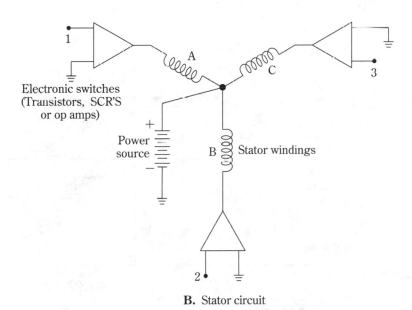

B. Stator circuit

5-4 An experimental brushless dc motor.

connections between terminals 7 and 12 on these IC modules render their operation similar to that of a differential op amp with sufficient output capability to reliably trigger triacs. When the incoming logic signal is "high," the output from the three IC modules are sustained dc drive voltages that trigger their respective logic triacs. In

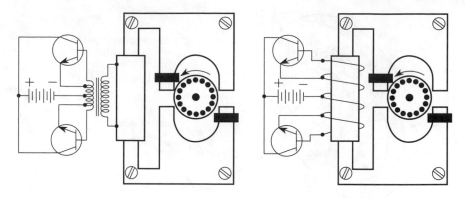

Inverter   Shaded-pole motor

A. Separate inverter and shaded-pole motor.

B. Integral implementation of inverter and shaded-pole motor.

5-5 Other types of brushless dc motors.

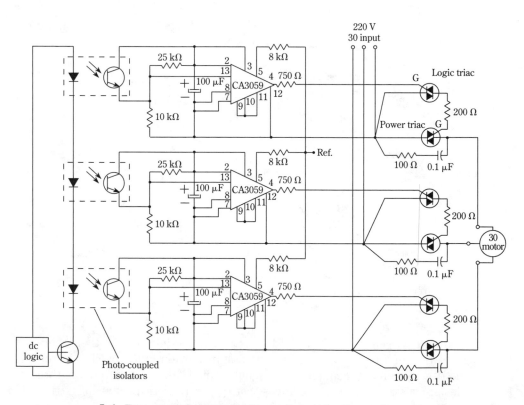

5-6 Triac control circuit for three-phase induction motors. RCA.

turn, the logic triacs gate the motor-controlling power triacs. As is the nature of all thyristors, the triacs automatically commutate to their off states when their ac currents approach zero in the progressing sine waves. Providing that the input logic remains high, each triac is retriggered twice per cycle of its phase voltage.

Inasmuch as a single logic signal controls the entire three-phase motor circuit, it is easy to associate various types of sensors with the input circuit so that the motor can be automatically turned on or off at certain temperatures or under other conditions. Similarly, timing circuits can be readily arranged to initiate or terminate motor operation when desired.

Variable-frequency inverter for speed control of a three-phase motor

The three-phase induction motor is probably the most important prime mover for integral-horsepower industrial applications. For the same weight, the rating of a single-phase induction motor is only about 60 percent that of the polyphase machine. Both efficiency and power factor are lower in single-phase motors. However, the three-phase induction motor has often exasperated the application engineer because of its resistance to speed control. Some extension of speed control is provided by the wound-rotor types, but the range is not competitive with that available from dc motors. "If only we could vary the applied frequency," was a remark often made in the era before solid state. Of course, in some implementations, the frequency was made variable by the use of another motor/alternator set. Such strategy is obviously not economical.

The design of an adjustable frequency, three-phase supply is not difficult if a compromise is made with regard to waveshape. Although a square wave would result in high eddy current and hysteresis losses, it is still not necessary to synthesize a true sine wave. A stepped waveform consisting of six segments can be created by mixing logic pulses. Such a waveform will have relatively low third-harmonic energy—the chief culprit in eddy current and hysteresis dissipation. The use of digital logic provides great simplification because the IC modules displace complex discrete circuits in all three phases.

Another aspect of variable-frequency control is the necessity of changing the motor voltage by the same percentage as the frequency change. In small motors, this basic requirement can be circumvented by inserting resistances in each motor lead. However, such a technique would not be allowable with large motors unless the speed variation were restricted. The use of such resistance also degrades the speed regulation of the motor.

Figure 5-7 is a block diagram of a variable-frequency inverter for speed control of a 10-hp three-phase induction motor. Six power channels provide the required sequence of voltage steps needed to develop the quasi-sine wave. This sequence is repeated at 120° intervals in order to supply three-phase energy to the motor.

The three-phase wave synthesis imparted by the digital-logic modules commences with the CL_1 pulse train. The subsequent modifications made to these pulses, together with various combining techniques, are illustrated in the logic-timing sequence shown in Fig. 5-8. Notice that a basic three-phase wave already exits with respect to waves A, B, and C. The circuit points from which these waves are de-

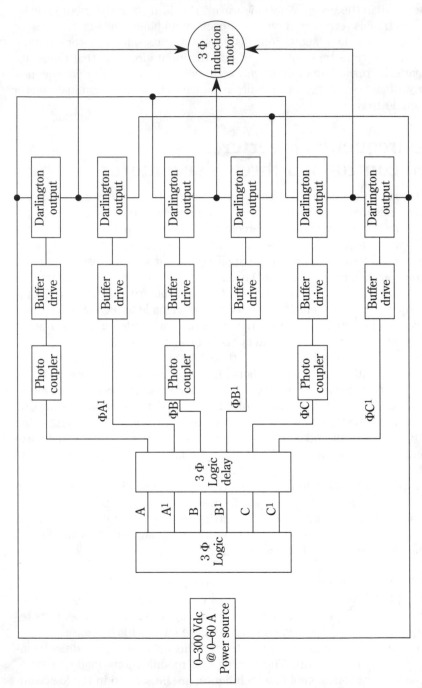

5-7   Block diagram of a variable-frequency inverter for speed control of 3-phase motors.   Delco Electronics.

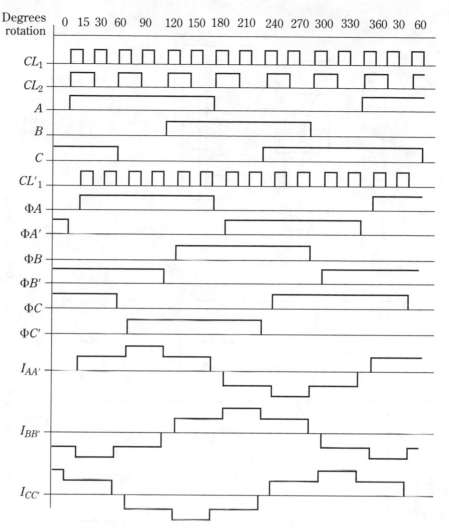

5-8 Logic timing sequence for variable-frequency inverter. Doloo Electronics

rived in the CMOS logic are designated in Fig. 5-9. Flip-flops FF102, FF103, and FF104 are connected as a three-bit shift register. Notice that the shift register provides complement waves A', B', and C' as well. These complement waves are used in the building-block process to modify the original ABC three-phase waves. One desired modification is a reduction in the half-wave duty cycle to 165° from the normal 180°. As mentioned previously, this prevents simultaneous conduction of power-output stages that are alternating their conduction states. The other desired modification is a very rough simulation of a sine wave. The CL_1 and CL_2 waves are also used here. (Although the stepped waves $I_{AA'}$, $I_{BB'}$, and $I_{CC'}$, ultimately delivered to the motor could never actually be mistaken for sine waves, they are nearly as suitable for operation of the motor as a true sine wave. Their salient features are the ease with which they are produced in logic circuits and their reasonably low third-harmonic content.

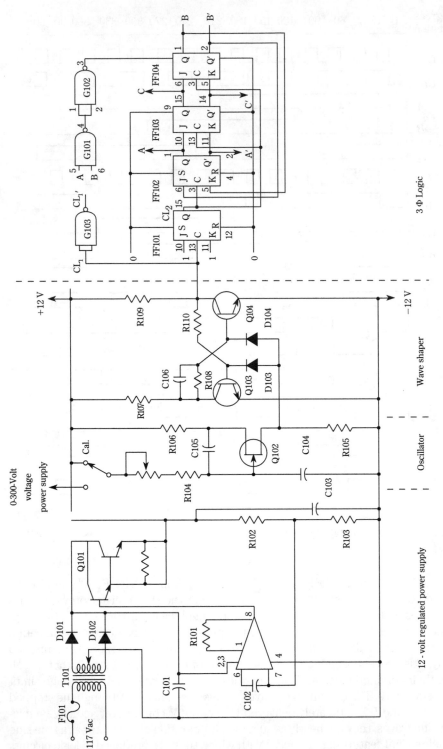

5-9 Voltage-controlled oscillator and 3-phase logic circuit. Delco Electronics.

The NAND gates, G101 and G102, inhibit forbidden-state operation of the shift register. Flip-flop FF101 halves the pulse rate received from the waveshaper, thereby producing wave CL_2, which clocks the modules in the shift register. Finally, an additional clock signal is produced by gate G103. This clock signal, CL_1, is not utilized in the generation of the basic three-phase waves. Rather, it is used to clock flip-flops FF411, FF421, FF431, FF451, and FF461, which appear in the schematic diagram of Fig. 5-10. These flip-flops constitute the 3 Φ Logic Delay shown in the block diagram of Fig. 5-7. This logic operation imparts the zero-voltage step to each wave in the three-phase sequence.

The basic frequency generated by the voltage-controlled oscillator is 24 times higher than that ultimately delivered to the motor. This is a useful technique of logic synthesis. A low-frequency wave can be constructed from higher-frequency "building blocks." Although suggestive of Fourier synthesis, the method is simpler—rectangular pulses are manipulated and combined to produce the desired waveform. The pulse synthesis is accomplished in the 3 Φ Logic function block of Fig. 5-7.

An important aspect of the waveshape produced by the logic circuitry is the zero-voltage step. As a consequence of this step, the duration of half cycles applied to the motor is 165°, rather than the usual 180°. This prevents conduction overlap between power-output stages that are turning off and those that are turning on. The function block that produces this zero-voltage step is the 3 Φ Logic Delay in Fig. 5-7. Photocouplers provide electrical isolation between the three-phase logic circuitry and the motor-drive circuits.

The schematic diagram of Fig. 5-9 shows the voltage-controlled oscillator and the basic three-phase logic of the inverter. Note that only the charging circuit for Q102, the unijunction transistor, connects to the external 0-300-volt variable-voltage supply. This allows adjustment of the charging rate of capacitor C104, and therefore of the pulsing rate of the emitter-base breakdown phenomenon of Q102. As a consequence, the generated frequency and the voltage applied to the motor go up and down together. The near-constant frequency produced by most unijunction oscillators is due to the fact that the emitter and base circuits connect to the same power source. In this circuit, the interbase voltage is fixed at 12 volts by the internal regulated power supply.

Following the unijunction voltage-controlled oscillator is a driven multivibrator involving transistors Q103 and Q104. This stage processes the oscillator signal to a suitable level and form for actuation of the 3 Φ logic modules. The output from Q104 consists of a periodic wave train of rectangular pulses CL_1. These pulses are at half the frequency of those initially generated by the unijunction stage.

Figure 5-10 is the schematic diagram of the three-phase inverter. This circuitry supplies power to the motor when driven by the signals generated in three-phase logic circuit previously described. Included in this schematic are the delay logic for imparting the zero-voltage step in the motor waveforms, and the photocouplers, which provide electrical isolation between low-level logic signals and the power circuits.

The circuit in Fig. 5-10 is simpler than it looks from an initial inspection. There is much repetition in the circuit configurations. Not only are all circuits for each of the three phases identical, but each of the six power channels contain five Darlington output amplifiers connected in parallel. The Darlington amplifiers are type-DTS,

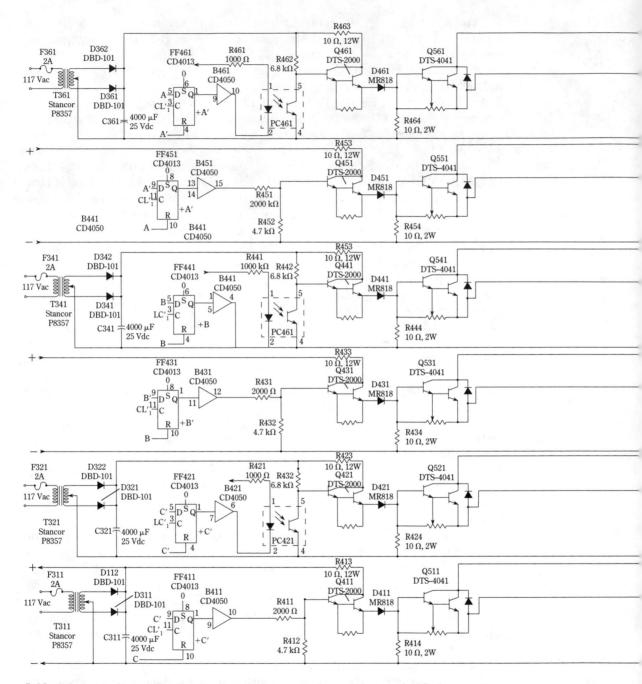

5-10 Schematic diagram of variable-frequency inverter for speed control of 3-phase motors.
Delco Electronics.

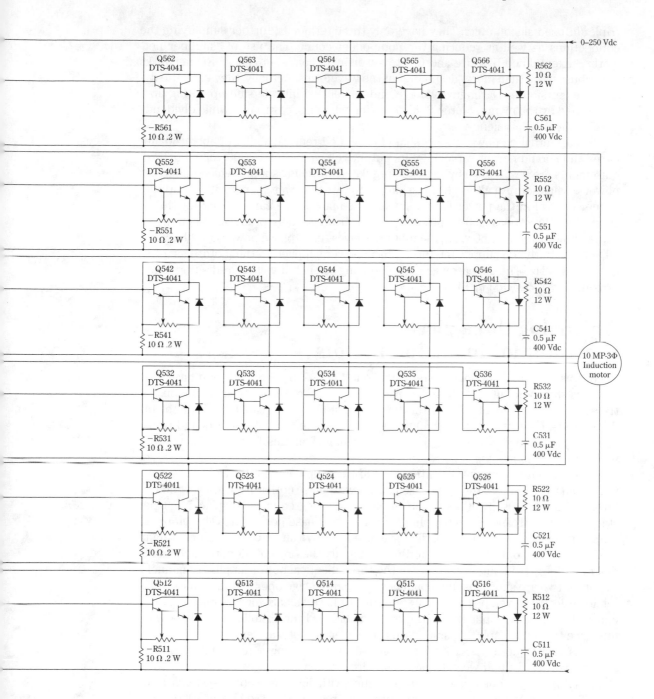

triple-diffused silicon units. They have only three terminals, further simplifying the connections as well as reducing the component count. The use of "snubber networks," exemplified by R512 and C511, is worthy of comment. These RC networks contribute more than mere "despiking" of the waveforms. Actually, they keep the dynamic excursion of Darlington voltage and current within the safe operating area prescribed by the manufacturer. Such load-line "shaping" pays worthwhile dividends in reliability and efficiency.

The external dc power supply should have a current capability of about 60 amperes and should permit variation of its output voltage from 0 to 300 volts. Such requirements are best served by a voltage-regulated switching supply, which minimizes power dissipation. An alternative is a simple full-wave power supply with autotransformer control in the power line. A single-phase power line can be used for the dc supply, but three-phase, 60-Hz power would probably be more practical. When the dc supply is adjusted for 250 volts, the motor receives three-phase power at 60 Hz. With 20 volts from the dc supply, the three-phase power supplied to the motor has a frequency of 5 Hz. The speed range corresponding to such a voltage reduction is approximately 1750 RPM to 145 RPM.

An analog three-phase generator for induction motors

In contrast to the digital logic system shown in Fig. 5-9, an analog three-phase generator is shown in Fig. 5-11. This circuit produces square waves because the output transistors are driven into their saturation regions. The power capability is on the order of 150–250 watts, total. The 400-Hz frequency is fixed. Both motors and transformers have significant size and weight advantages at this frequency, compared with 60-Hz components.

The oscillator is essentially a three-stage RC-coupled amplifier with a feedback path from the output to the input. Oscillation takes place at that frequency, which allows a total phase shift of 360°. This implies a 120-degree phase displacement per stage, which corresponds to the phasing of a three-phase power line. Therefore, an output from each stage of the oscillator is used to drive an amplifier, culminating in push-pull power output stages. In order to simplify the presentation, the schematic diagram of Fig. 5-11 shows only *one* of the three identical power-amplifier channels; the other two are shown as function blocks. The first stage operates as a phase splitter in order to develop the 180-degree out-of-phase signals for the push-pull circuitry. In this way, input transformers are avoided. Emitter-follower drivers then interface this phase-splitting stage with the power-output transistors.

This technique is excellent for applications demanding a respectable ratio of horsepower to size and weight, such as in aircraft, space vehicle, and marine work. Control flexibility is not as good as that obtained with digital polyphase systems, however. To get a speed-control range, substitute a three-ganged assembly of 1000-Ω potentiometers for the 820-Ω resistors. The wiper of each potentiometer would then connect to the coupling capacitor feeding the base of the transistor associated with that potentiometer. The idea is to vary the time constants without changing the tran-

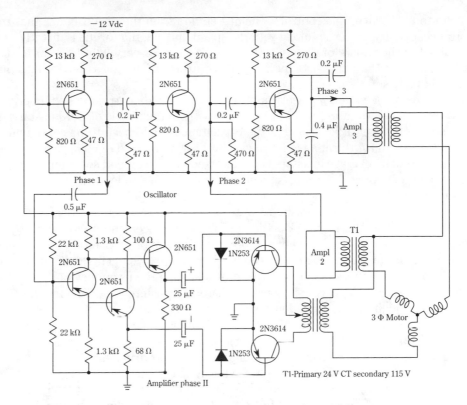

5-11 Analog three-phase generator. Motorola Semiconductor Products, Inc.

sistor bias networks. This, alone, will provide higher frequencies and greater motor speeds. If lower speeds are also desired, the 0.2-µF coupling capacitors should be increased in size. If the oscillator frequency is reduced more than about 15 percent of the nominal 400 Hz, an arrangement should be made to simultaneously reduce the voltages applied to the motor.

Sine-wave operation of this system could be achieved by increasing the value of the 47-ohm emitter resistances in the three oscillator stages. However, efficiency of the power-output transistors would then be reduced from approximately 90 percent to approximately 70 percent.

Logic-circuit speed controller for a permanent-capacitor, split-phase motor

The permanent-capacitor, split-phase motor is particularly attractive for the application of electronic control. This motor has electrical symmetry and operates smoothly from a two-phase power source. One way to obtain speed adjustment is to vary the *applied voltage*. This scheme exploits the slip characteristic of the motor. Although the permanent-capacitor, split-phase motor permits an inordinate amount of slip to take place, this control technique suffers from the disadvantage that the torque capability decreases along with the speed. Indeed, the torque is directly proportional to the square of the applied voltage; except for very light loads or for fan and blower applications, this type of speed control is often not as practical as one might initially suppose.

A *variable-frequency* power supply is a better speed-control method. In principle, a wide speed variation should be possible without the plague of "torque-starved" operation. However, it is clear that the capacitance would have to be continuously variable *along with* the applied frequency. At best, some kind of compromise could be made so that several capacitors could be selected by a tapped switching arrangement. This, of course, discourages the implementation of a wide speed-control range in such a system.

A better approach is to dispense with the capacitor(s) and to utilize logic circuits to establish the required 90-degree phase displacement *regardless of frequency*. This is done in the scheme represented by the block diagram of Fig. 5-12. Linear, digital, and optoelectronic techniques are employed. This is an open-loop system and the motor speed is controlled by adjusting the frequency of the oscillator. Because the optical couplers have unilateral transference characteristics, together with high voltage isolation between their input and output circuits, motor transients are clearly isolated from the logic circuits (such transients could arise from sudden load variations on the motor).

The schematic diagram of this controller circuit is shown in Fig. 5-13. The use of IC modules renders the actual circuit only slightly more complex than the block diagram.

In the ensuing discussion of the overall circuit, it is suggested that reference also be made to the waveform diagram of Fig. 5-14. Circuit operation is simpler than one might assume from these numerous waveforms, for many are paired or are image-related waveforms due to the biphase and push-pull aspects of the configuration.

The schematic diagram in Fig. 5-13 shows a free-running relaxation oscillator using a unijunction transistor, Q1. The oscillator frequency, and therefore the speed of the motor, is adjustable by the 500-kΩ potentiometer in its emitter circuit. This oscillator has a frequency range of approximately 40 Hz to 1200 Hz, but subsequent logic operations divide the frequency down to the 10-Hz to 300-Hz range. This corresponds to a speed range of 300 RPM to 9000 RPM if a two-pole motor is used.

Transistors Q2 and Q3 are essentially waveshapers. They process the output from the UJT oscillator to a suitable waveform for actuating the set-reset input circuits of the "X" flip-flop within the MC688 IC. This flip-flop is the R-S type—its operation depends on the input levels and their duration, rather than their rise and fall times.

The $\overline{Q}$ output of the "X" flip-flop is designated as $\overline{X}$ in both the schematic and the waveform diagrams. The $\overline{X}$ waveform clocks the "A" flip-flop, which is also within the MC688 IC. The "A" flip-flop performs as a divide-by-two toggle and provides two clock signals, which are 180° out of phase. These clock signals are applied to the "B" and "C" flip-flops.

The "B" and "C" flip-flops toggle on the negative transitions of their input clock signals. At the same time, these flip-flops divide their input clock rates by two. A comparison of output wave trains from flip-flops "B" and "C" shows a phase dis-

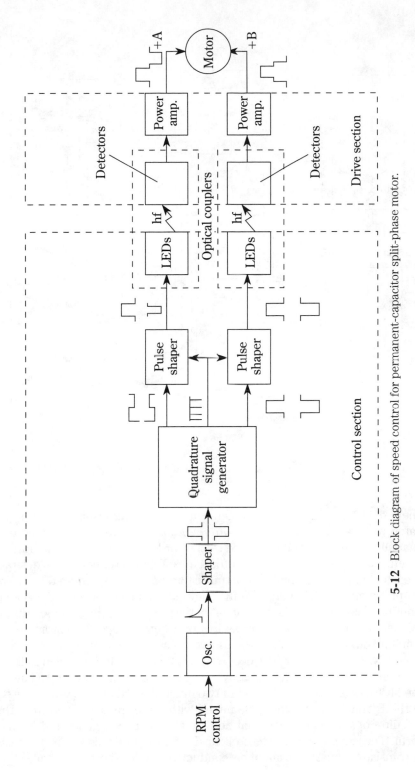

5-12 Block diagram of speed control for permanent-capacitor split-phase motor.

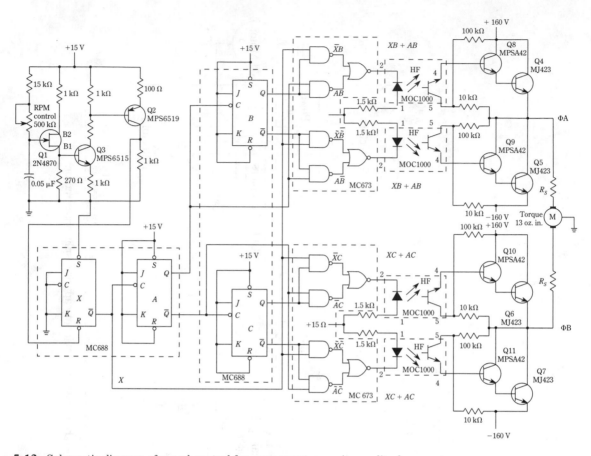

5-13 Schematic diagram of speed control for permanent-capacitor, split-phase motor. Motorola Semiconductor Products, Inc.

placement of 90°. Significantly, this quadrature phase relationship—contrary to that associated with capacitor circuits—is *independent of frequency*. It would now appear that all that need be accomplished is a power-level boost of the quadrature output waveforms. Although this might suffice, an additional refinement is desirable.

In each channel, push-pull power transistors provide the required ac power for the two-phase motor. The operating efficiency of such push-pull amplifiers can be increased if provision is made for turning off both transistors during the polarity transitions. Otherwise, there will be high dissipation when the two transistors are forced into simultaneous conduction during crossover, because of their inability to turn off instantaneously. The reason for such occurrence is that turn-off time is generally slower than turn-on time. Cutoff during crossover is produced in the following manner:

The MC673 ICs contain a dual set of two input of NAND-NOR gates. These gates enable the B and C waveforms to be mixed with pulse trains derived from the "X" and "A" flip-flops in such a way that zero-voltage steps are generated in the final waveform. These zero-voltage steps occur during the intervals when the push-pull output transistors must alternate their conductive states. The zero-voltage steps are

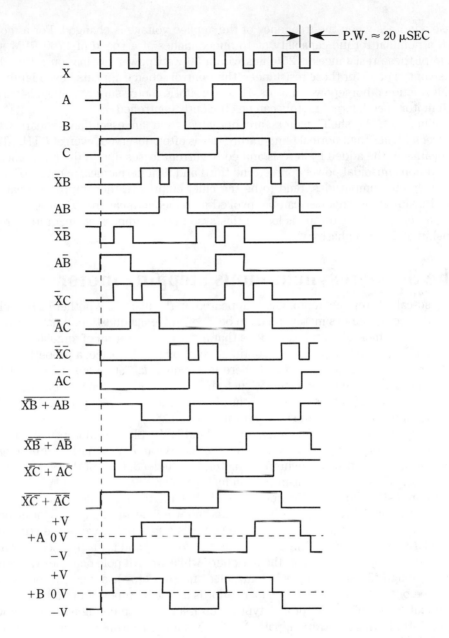

5-14 Waveform diagram for speed-control circuit in Fig. 5-13. Motorola Semiconductor Products, Inc.

clearly seen in the ΦA and ΦB motor-drive waveforms. These "rest periods" for the power transistors are actually only about twenty microseconds in duration; they represent a very minor distortion of the essentially square waveforms.

The resistances R_s, in series with the stator windings of the motor, limit its current. Such limiting is necessary to compensate for the change in the reactance of these windings when the frequency of the applied voltage is changed. For a motor with a nominal torque capability of 13 ounce-inches at a speed of 1700 RPM, the value of these resistances is 25 ohms with a 50-watt power dissipation rating. Because of the need for these resistances, this control scheme appears best adapted to small fractional-horsepower motors. However, if the motor supply voltage is varied with motor speed, larger motors can be efficiently controlled.

When "off-the-shelf" motors are operated from an essentially square-wave source, a higher than normal temperature rise is often observed, even at 60 Hz. This is because of the added hysteresis and eddy current losses due to the harmonics in such a nonsinusoidal power source. The third and fifth harmonics, because of their relatively high amplitudes, tend to be the chief culprits. Either improved heat removal or lower horsepower can be invoked as remedial measures. On the positive side, no efficiency or torque is lost in this system from imperfect quadrature relationship of the two phases.

The Slo-Syn synchronous stepping motor

Many so-called stepping motors have appeared on the market. Whether or not all of these devices qualify as motors tends to be a controversial question. Whatever their form, the common operational mode is that a discrete quantum of angular rotation occurs in response to a pulse. For continual rotation to take place, a properly coded pulse train must be applied so that there is a sequential "stepping" motion of the shaft. In essence, the pulse coding is such that a *rotating field* is produced. One of the salient features of these devices is the ability to repeat positional information. No feedback loop is required for such performance.

A particularly interesting device of this kind is the Slo-Syn motor made by the Superior Electric Company. Significantly, the Slo-Syn motor is marketed both as a *stepping motor* and as a synchronous motor. It is, indeed, both of these. Moreover, it is described as a permanent-magnet inductor motor.

Figure 5-15 is an exploded view of the Slo-Syn motor. It is evident that there are no commutator or brushes, no slip rings, and no windings on the rotor. Not evident is the nature of the stator windings. The stator is a four-pole, two-phase structure with a large number of "teeth." The rotor is also *toothed* and is magnetized so that a south pole occupies one-half of the periphery while a north pole occupies the other half. Although the stator has only four poles and the rotor has only two poles, the presence of the teeth on both members affords a large number of opportunities for positional "lock up" of the rotor. A typical rotor-stator tooth configuration is shown in Fig. 5-16. In this illustration, poles N1, S3, N5, and S7 form one phase; poles N2, S4, N6, and S8 form the other phase.

An insight into operation from both pulsed and sinusoidal power is provided by the arrangement shown in Fig. 5-17. In essence, the potentiometer constitutes a two-phase generator. One complete revolution of this potentiometer advances the angular position of the rotor by one tooth, in the same way that one cycle from a conventional ac supply would. With a rotor having 100 teeth (two 50-tooth sections), 100 revolutions of the potentiometer would be required for one complete revolution of the motor shaft. The positions of the potentiometer and motor always

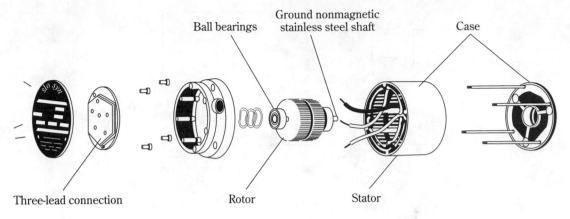

Ball bearings

Ground nonmagnetic
stainless steel shaft

Case

Three-lead connection

Rotor

Stator

5-15 Exploded view of the Slo-Syn motor. Superior Electric Co.

5-16 The rotor-stator tooth configuration for a 72-RPM, 60-Hz Slo-Syn motor. Superior
Electric Co.

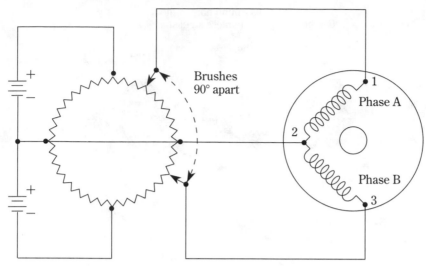

5-17 An arrangement for the "slow-motion" rotation of the Slo-Syn motor. Superior Electric Co.

remain synchronously related—a given angular position of the motor can always be "recaptured" by turning the potentiometer back to its total angular displacement from "start" that previously corresponded to that motor position.

The Slo-Syn motor will operate from a single-phase line when a suitable phase-shifting network is provided, such as that depicted in Fig. 5-18A. The basic idea, of course, is to make the motor stator "see" a two-phase supply. The motor then operates at subsynchronous speed, which is 72 RPM with a 60-Hz line when the rotor has 100 teeth. The fact that the stator has four poles is not the significant speed-determining factor, as it is with conventional synchronous motors. The 100-tooth rotor is equivalent to a 100-pole rotor, and the motor operates as if the stator also had this number of poles. The subsynchronous principle has been known for a long time and has previously been utilized in timing applications where constant speed was necessary and where it was desired to avoid complex, torque-consuming gear trains for speed reduction.

In the event of any appreciable change in frequency, either R or C in the phase-shift network must be changed. A compromise network is feasible for use with both 50- and 60-Hz lines. Additionally, when a constant-torque output is desired, the voltage applied from the line must be varied as shown in Fig. 5-19 because the winding impedance of the stator is a function of frequency. If constant-torque output is not of primary interest, it is conceivable that a range of speed variation can be obtained with fixed voltage. In such a case, the low-speed operation limit is imposed by temperature-rise considerations, because greater current will be carried by the stator windings at low frequencies. A favorable factor here is that starting, running, and stall currents are, for all practical purposes, the same.

A more convenient way to obtain a wide range of speed adjustment is shown in Fig. 5-18B. In this circuit, a two-phase source of power is used. No phase-shift network is needed. Applied voltage must still be increased with frequency if constant-torque output is desired.

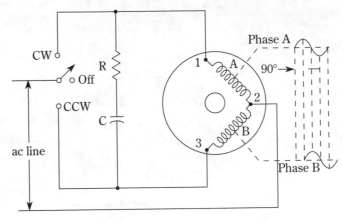

A. Operation from a single-phase power line.

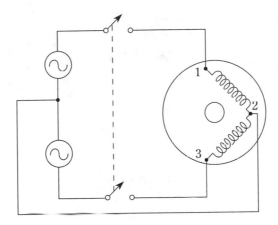

B. Operation from a two-phase power source.

5-18 ac operation of the Slo-Syn motor. Superior Electric Co.

Unlike various other timing and positioning motors, the Slo-Syn motor is capable of delivering considerable mechanical power—the Slo-Syn is made in frame sizes to accommodate torque outputs from 25 to 1800 ounce-inches.

The switching logic for sequential stepping of the Slo-Syn motor is shown in Fig. 5-20. These motors are also made with bifilar windings so that a single-ended power supply can be used. The stator retains its two-phase, four-pole winding format, and the basic operation of the motor is unchanged. Figure 5-21 illustrates the way logic is applied for sequential stepping of the bifilar-wound Slo-Syn motor.

Dynamic braking techniques for ac motors

The traditional way of shortening the coasting time of a *commutator* motor following turn-off is to consume current from the still-rotating armature. The machine

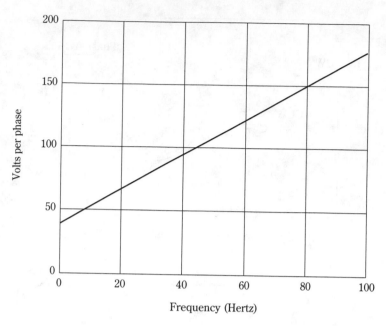

5-19 Required voltage when constant-torque output is desired from a Slo-Syn motor. Superior Electric Co.

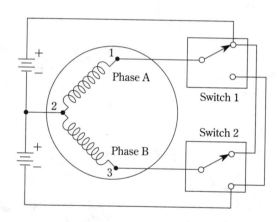

5-20 Switching logic for sequential stepping of the Slo-Syn motor. Superior Electric Co.

Switching sequence *		
Step	Switch 1	Switch 2
1	+	+
2	−	+
3	−	−
4	+	−
1	+	+
* To reverse direction, read chart up from bottom		

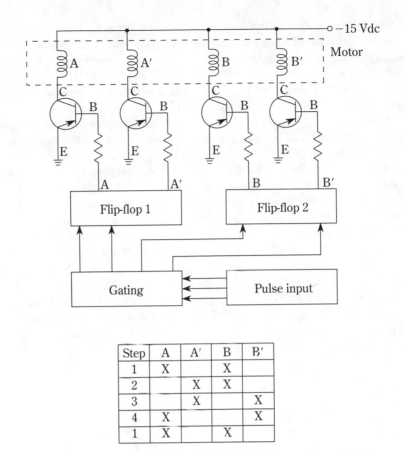

5-21 Sequential stepping-logic for bifilar Slo-Syn motor and single-ended power supply. Superior Electric Co.

Step	A	A′	B	B′
1	X		X	
2		X	X	
3		X		X
4	X			X
1	X		X	

then becomes a generator, and the electrical energy it delivers is derived from the kinetic energy stored in its rotating members. Rotation is therefore quickly deccelerated, and the motor is brought to a standstill in much shorter time than it would be from windage and bearing friction (see Fig. 4-24). *Noncommutator motors*, however, are not suitable for this method of dynamic braking.

Two-phase and three-phase induction motors can be dynamically braked by "plugging," that is, by attempting reverse rotation. This is accomplished in these polyphase motors by transposing power-line connections to one phase of the motors stator winding. Because the permanent-capacitor, split-phase motor is, in essence, a two-phase machine, it too can be dynamically braked by the ordinary reversing technique. Of course, all motors braked by attempted reversal must be disconnected from the power line before actual reverse rotation occurs. The resistance-start induction motor, or what is loosely referred to as a "split-phase" motor is not readily braked by transposition of its windings, nor is the capacitor-start induction motor. The basic idea is that those motors, which ordinarily should not be reversed until the shaft is at rest, should not be subjected to the "plugging" techniques.

All ac motors with squirrel-cage rotors can be dynamically braked by injecting dc into one, or more, of their windings following disconnect from the ac power line. In essence, such machines then become separately excited dc generators with shorted armatures. Such braking can be very effective, and there is no tendency for reverse rotation. Immediately after standstill is achieved, the dc should be disconnected to prevent heating of the stator winding(s).

A more sophisticated method for accomplishing such braking is to dump the electric energy stored in a capacitor into one or more stator windings of squirrel-cage motors. The circuit shown in Fig. 5-22 is applicable to shaded-pole motors and all other ac motors with squirrel-cage rotors, providing they are not too large—¼ hp is perhaps a practical limit because of the inordinate capacitor size required for larger machines. Depending on the motor size and the deceleration desired, capacitors up to hundreds of microfarads are feasible for many braking requirements. Diode D1 can be a silicon rectifier with a current rating of several amperes. Resistor R1 can be a two-watt composition resistor of about 50 kΩ or so.

Synchronous motors can usually be braked by the application of dc to the stator, but the decelerating mechanism is somewhat different from that of induction motors. Empirical investigation is often required for optimum results. A possible advantage attending dynamic braking of synchronous motors is their ability to lock in position once brought to a standstill. This "holding" behavior is most pronounced in the *reluctance* synchronous motor.

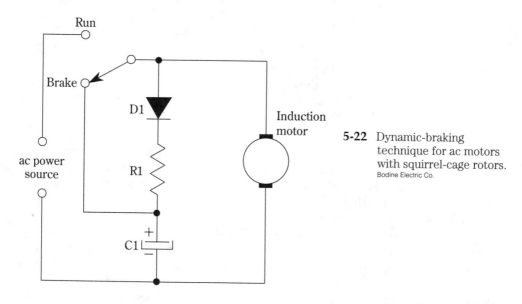

5-22 Dynamic-braking technique for ac motors with squirrel-cage rotors.
Bodine Electric Co.

The electronic regulation
of the automotive alternator

With the aid of solid-state devices, the automotive industry has quite successfully overcome an irritating reliability and maintenance problem that had plagued the family car for several decades. Although the dc generator formerly used to keep the battery charged incorporated some clever design principles, it was often subject to trouble. This stemmed primarily from the need for the commutator and brush system to carry the full charging current. The overall effects of arcing, wear, and dirt rendered this vital engine accessory a candidate for failure—and all too often, when least anticipated. Also, the unreliability of the automotive electrical system was further reinforced by the electromechanical *voltage regulator* associated with the generator.

Modern cars use three-phase alternators with self-contained silicon-diode rectifiers. These machines employ slip rings to conduct current to the rotating field. However, a slip ring is a much simpler device than a commutator. Moreover, the field current is a small fraction of the charging current often supplied to the battery. Therefore, the slip rings and brushes require infrequent maintenance compared with the commutator and brush assembly of the old dc generators. The electromechanical voltage regulator did not immediately become obsolete, but the trend is now clearly toward the electronic voltage regulator.

A typical electronic voltage regulator is shown in Fig. 5-23. This circuit senses the battery voltage and causes the alternator to deliver either full charge or none at all. This operating mode simulates that of the electromechanical regulator. Whereas the electromechanical regulator might turn the alternator field current off and on at a 100-Hz rate when the charge state of the battery was marginal, the electronic regulator switches on and off at a rate in the vicinity of 1000 Hz. This in itself is of no particular advantage. The important feature is that there is no mechanical wear in the solid-state type voltage regulator.

In the circuit of Fig. 5-23, the input transistor conducts only when battery voltage becomes high enough to break down the zener diode in its base lead. When this happens, the Darlington output stage is turned off and the field winding of the alter-

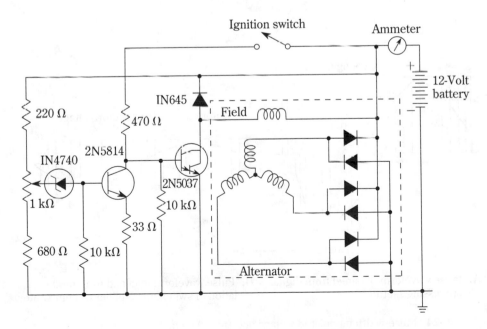

5-23 Electronic voltage regulator for automobile alternators.

nator is deprived of current. Depending on the battery and its load, the field circuit will be energized and opened at widely varying duty cycles. This type of regulator is more closely related to the switching-type power supply rather than to the "linear" voltage-regulated supply.

The circuit of Fig. 5-23 is not critical—many types of power transistors will be found suitable. It is particularly adapted for use in mobile radio and in standby power systems that use the automotive alternator.

The view ahead

In bringing to a close this investigation of electronic controls for electric motors, it is appropriate to discuss some trends likely to assume increasing importance. Rather than speculate over the possibilities of applying the magnetic monopole or harnessing gravitational force, I prefer to deal with the devices, techniques, and systems that are presently realizable.

Pulse-width modulation systems for ac motors

This variable-frequency technique fabricates quasi-sine waves from weighted pulses of constant amplitude. The waveshapes depicted in Fig. 5-24 are relevant. Because of the filtering or integrating action of the motor inductance, the *motor current* is a satisfactory approximation of sine wave. One method of producing pulse-width modulation is to deliver properly timed gate trigger pulses to an SCR circuit such as that shown in Fig. 5-25. This is somewhat easier than might initially be expected. Ordinarily, the gate control of an SCR is lost once triggering has occurred. However, in this circuit a pair of SCRs, such as Q1 and Q2, operate like an SPDT toggle-switch.

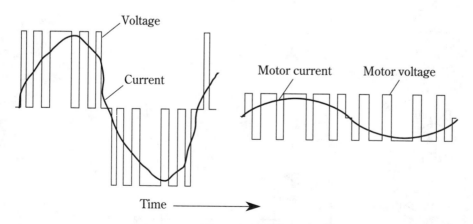

A. Pulse waveform obtained from digital processing method.

B. Pulse waveform obtained from modulation of switching amplifier or comparator.

5-24   Pulse-width modulated voltage and smoothed motor-current waveform.

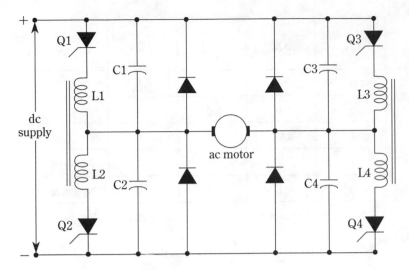

5-25 Simplified schematic of the thyristor pulse-width modulator.

Assume that Q1 is in its on state. Then, by triggering Q2 to its on state, Q1 is automatically commutated to its off state by the polarity of the induced voltage across L1. Similarly, anytime Q1 is triggered on, conduction in Q2 is extinguished due to L2. The same relationships exist with regard to Q3 and Q4. The truth table for this behavior is shown in Fig. 5-26. The sequential triggering pulses required to produce the pulse-width-modulated waveform can be generated by various digital logic systems, and can include programmable ROMs, multiplexers, and even microprocessors.

Pulse-width modulation can also be produced by linear/digital arrangements. For example, the method shown in Fig. 5-27 makes use of two LM111 comparator ICs. These are connected to be self-oscillating, thereby providing a "carrier" frequency. A lower-frequency sine wave impressed at the input of the arrangement results in a pulse-modulated output wave, such as the one illustrated in Fig. 5-24B. Power can be boosted by means of bipolar-transistor output stages.

The cycloconverter

The cycloconversion principle provides another means of efficiently varying the frequency of the power applied to an ac motor. A cycloconverter circuit is shown in Fig. 5-28A. The equivalent circuit in Fig. 5-28B is useful in analyzing the operation of cycloconverters. Two frequencies are applied to the system. One of these is f_1, which is often obtained from the ac power line. Frequency f_1, can be considered a "carrier" modulated by the second frequency, f_2. The output frequency is also at f_2 and does not appear to have a waveshape desirable for motor operation. However, the inductive reactance of ac motors is sufficient to convert the current wave into a fairly good sinusoid. This is particularly true when operation is from a three-phase line. A simplified schematic diagram of a three-phase to a single-phase cycloconverter is shown in Fig. 2-28C. The motor waveforms corresponding to a frequency-reduction ratio of three are illustrated in Fig. 5-28D. The ratio of line frequency to modulation fre-

Q1	Q2	Q3	Q4	Motor voltage
0	0	0	0	0
0	1	0	1	0
1	0	1	0	0
1	0	0	1	Positive
0	1	1	0	Negative

5-26 Truth table for pulse-width modulator shown in Fig. 5-25.

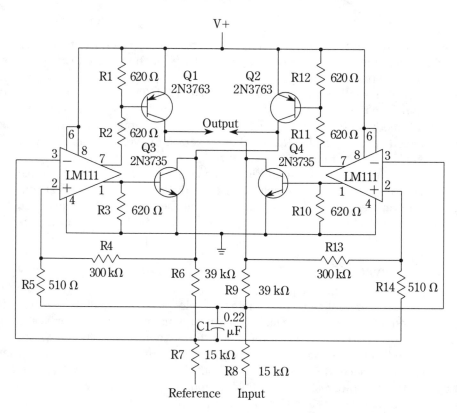

5-27 Integrated-circuit system for producing pulse-width modulated waves. National Semiconductor Corp.

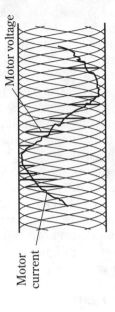

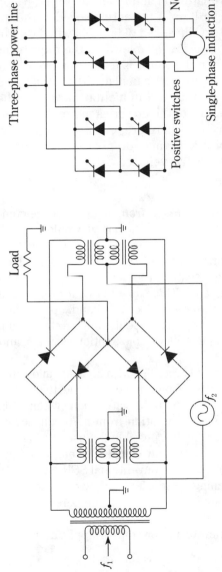

C. A three phase to single phase cycloconverter (gating logic not shown).

D. The voltage and motor-current waveforms for a 400-Hz power line and a 135-Hz modulating (output) wave.

A. A single phase to single phase cycloconverter.

B. Equivalent circuit to (A).

5-28 The cycloconverter.

quency can be as low as two in polyphase systems—below that, serious deterioration of the output wave takes place. There is no high limit to the frequency-reduction ratio, however. Indeed, the cycloconverter performance improves because the output wave is then fabricated from a large number of small elements.

With the cycloconverter, an induction motor can be operated from zero speed to one-third, and in some cases one-half, the speed that would otherwise be obtained from the power-line frequency. In addition to providing efficient and continuous speed control, the cycloconverter enables an induction motor to develop maximum torque at slow speeds. It is fortunate that induction motors develop maximum torque at a certain slip speed, regardless of the actual speed of the rotor. Thus, if maximum torque corresponds to a slip speed of 50 RPM when the rotor is turning at 1740 RPM, the maximum torque is available when the rotor turns at a much slower speed. This is achieved when the stator is fed with a frequency that makes the difference between the rotating magnetic field and the rotor speed equal to 50 RPM. The significance of this is that the induction motor can be caused to simulate the powerful tractional effort of the dc series motor, but without its commutation problems and maintenance. Therefore, even though the cycloconverter technique has been known for a long time, do not be surprised to find that it has been rediscovered for use in electric vehicles.

SCRs with gate turn-off capability

It is obvious that a more ideal control device would result from an appropriate combination of thyristor and transistor characteristics. In particular, it would be very useful to have available an SCR or triac that could be gated off as well as on. The semiconductor manufacturers have been working toward this objective for a long time. The General Electric silicon-controlled switch, the Motorola gate-control switch, and the RCA GTO device are examples of SCRs capable of turn-off by the gate signal alone in an ac or dc circuit. Some of these devices can handle power levels much higher than usual "signal-level" thyristors. At present, they are beginning to see application in switching-type power supplies, electronic ignition systems, and in controls for fractional-horsepower motors.

Figure 5-29A illustrates an RCA GTO controlled rectifier with 50 watts allowable dissipation. Figure 5-29B and C show two turn-off techniques for the RCA GTO thyristor. Both circuits provide the required—70 volts to the gate when switch S1 is closed. Whereas the circuit of Fig. 5-29B requires operation from a 70-volt power supply, the addition of the inductor in Fig. 5-29C makes possible the use of a 35-volt power supply. Of course, an auxiliary 70-volt power supply can also be employed for turn-off. Turn-on is initiated by a positive pulse, as in a conventional SCR. Switch S1 can be a transistor, a thyristor, or other switching device. The family of devices exemplified by Fig. 5-29 can handle a nominal steady current of 15 amperes and can block up to 600 volts. Present development is aimed at 200-ampere, 1000-volt units. When such ratings are attained, the relevancy to motor-control applications can hardly be overestimated.

Motors designed for a wide range of control

For the most part, the electronic control techniques described in chapters 4 and 5 work reasonably well with off-the-shelf motors. However, the majority of these mo-

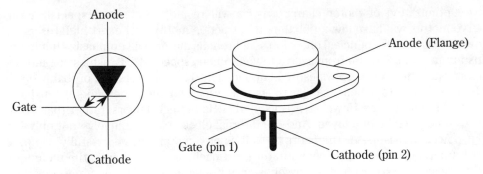

A. Symbol and appearance of device.

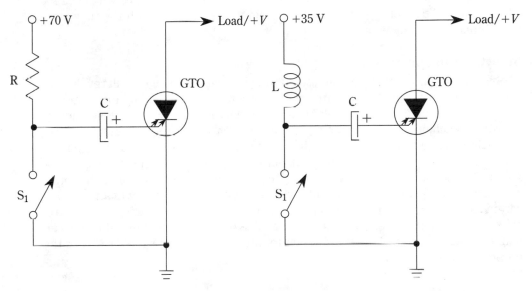

B. Gate turn-off technique using a 70-volt power supply.

C. Gate turn-off technique using a 35-volt power supply.

5-29 The RCA GTO silicon-controlled rectifier.

tors were not originally intended for wide ranges of control and might suffer in efficiency when operated at very low and very high speeds. This adversely affects other performance parameters such as torque, speed regulation, and acceleration, causing increased burden to be placed on the electronic control and drive systems. If these motors were designed to keep pace with the new control techniques, smoother and more reliable operation would result. Because solid-state technology is more flexible than motor technology, there will probably always be somewhat of a lag between the two. However, it is now widely realized that motors must be made with lower eddy current and hysteresis losses; with ability to withstand high acceleration, rapid braking, and abrupt reversal—and with superior insulation.

Optimization of motor characteristics will result in extended specialization; servo motors will have low rational inertia, traction motors will exert high torque at low speeds and frequencies, precision synchronous motors will emphasize minimum instantaneous deviation from average synchronous speed, etc. Brushless dc motors appear destined for more widespread application. These motors will probably operate according to a scheme similar to that described for the experimental brushless motor in this chapter. However, it is likely that Hall-effect sensors, rather than optoelectronics, will be employed. And the internal electronics will involve not only ICs but power transistors or thyristors as well. Such a motor will successfully compete with fractional-horsepower commutator-type machines. In principle, the technique is applicable to larger integral-horsepower motors, but the ac induction motor then looms as a formidable rival.

Because feedback systems will become increasingly popular, there will be a wider selection of useful packages, such as motors with built-in tachometers and encoders. Also, printed circuit motors and the newer hollow-rotor types will, because of their capabilities for high acceleration, team up increasingly with the new electronic control techniques.

Smart motor controls

Sophisticated control of electric motors, now in its beginning stages, will revolutionize industrial processes. Such control will incorporate computers and microprocessors, and will implement the popular concept of "automation." For example, motors that drive machine tools to produce three-dimensional contouring of metal will carry out their assigned tasks according to the directives stored in programmed memories. Moreover, many such "smart" motors will perform together. Allowances will be made for the individual and momentary cutting and milling requirements of each motorized "machinist," and all of the contoured objects will emerge identically shaped within close tolerances. The social and economic implications of such motor control are obviously great. Surely, a technology capable of multiplying material goods can help realize the universal quest for the better life. Its attainment simply requires the proper programming format.

6
Control applications for a variety of electric motors

THE PREVIOUS CHAPTERS DEALT WITH BASIC PRINCIPLES OF VARIOUS ELECTRIC MOTORS and with detailed circuit techniques for controlling them. Once insight has been attained with regard to the operation and performance of the different types of motors, it becomes feasible to treat the topic of electronic control in a more streamlined fashion. Accordingly, this chapter presents examples of electronic control for a variety of electric motors, but under the assumption that you no longer require a detailed "how-to" approach. This makes sense because individual performance requirements can be expected to differ widely. Also, it is generally true that the "classical" behavior of the various motors no longer prevails after applications of the electronic control.

Energy conservation via power factor control

Next to the electric lamp, the induction motor has contributed greatest to the modern lifestyle associated with the age of electricity. In particular, the single-phase type is the prime mover behind a wide variety of appliances, pumps, and mechanized tools. Until recently, relatively little has been done to alter the performance of induction motors. Indeed, its ability to maintain speed and torque in the face of both voltage and load variations has long been hailed as its most desirable feature. Overlooked, at least until the advent of society's "energy crunch," has been the lamentably poor efficiency of these motors at *light* loads. And prior to a clever electronic invention, the general notion was that virtually nothing could be done to alleviate this drawback.

The fully loaded induction motor consumes line current that is nearly in phase with the impressed voltage. With the power factor thereby close to unity, the operating efficiency of the motor is reasonably good. But, at light loads, the situation is quite different—for then there is considerable phase-lag between current and voltage. Because of this low power factor, the actual magnitude of the consumed current remains high. This causes much greater I^2R loss in both the motor and the line than would prevail if, somehow, better phase conditions could be restored. Of course, in such a situation, the magnitude of the current would decrease to comply with the low torque demand of the light load. In order to bring this desirable situation about, start by *sensing* the power factor of the line supplying the motor, and then vary an operating parameter that will affect the phase relationship.

Fortunately, it is only necessary to reduce applied voltage in order to improve the phase condition of the lightly loaded motor. Indeed, this approach enables restoration of the power factor to approximately its fully loaded value. In actual practice, this must be done *automatically* in such a manner that the motor always operates at a high power factor (voltage and current nearly in phase) for any load condition.

Figure 6-1 is a block diagram of a simple electronic system for maintaining a high power factor. By its use, the energy cost of running induction motors in some applications can be lowered by as much as 40 percent.

In the block diagram of Fig. 6-1, the several linear and digital functions can be implemented by a variety of op amps, transistors, diodes, and gates. One of the main objectives is to produce pulses, the width of which are proportional to the phase angle between line voltage and motor current. This important pulse train is shown in the waveform diagram of Fig. 6-2 as the logical format $\bar{E}.I + E.I'$ and is present in the block diagram at point "A." Note that Θ represents the phase angle between line voltage and motor current, and is the sensed parameter in the control system. The sensing of Θ and the subsequent production of width-modulated pulses at point "A" can be said to be the electronic heart of the control technique. This cause-and-effect relationship then actuates subsequent circuitry, which controls the *voltage* applied to the motor (the waveforms are idealized to simplify the basic discussion).

Initially, the voltage applied to the motor and the current consumed by the motor are applied to overdriven op amps in order to produce waveforms $\bar{E}$, and I, I'. The current waveforms are the consequence of sensing the voltage drop across the current-sense resistor, which is of very low value in order not to effect motor performance or develop appreciable dissipation. The logic gates then are employed to AND $\bar{E}$ with I, and E with I', thence ORing the result. This gives the desired pulses having widths proportional to the power-factor angle Θ.

Actually, a *pair* of op amps are used in both the voltage-squaring, and in the current-squaring functions inasmuch as *two* voltage waveforms and two current waveforms must be derived for presentation to the logic gates. From the waveform diagram of Fig. 6-2 the individual amplifiers of the respective pairs can be seen to develop 180° displaced outputs.

The next objective of this control system is to use the pulse train at "A" to vary the effective voltage applied to the motor. The basic idea is to automatically lower the effective motor voltage when the power factor is low, that is, when the motor is lightly loaded. This accomplishment will, in turn, greatly increase the light load power factor. This important result occurs because only then can the motor, operating with reduced voltage, accommodate its load. The improved power factor will be accompanied by reduced motor current. Thus, the control function merely deprives the motor of voltage when the sensed power factor is low (corresponding to light load). Thereafter, the motor adjusts its operation from low power factor and high

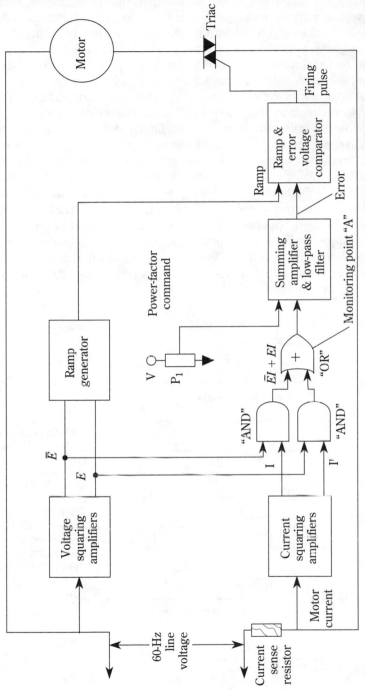

6-1 Block diagram of the power-factor control system for induction motors. Systems operates to lower *applied voltage* to motor at low power-factors. This, in turn, tends to correct the power factor and increase the efficiency of the motor. Monitoring point "A" is explained in the text.

current to the more efficient format of high power factor and low current. Let us now see *how* the motor voltage is reduced in response to the low power factor condition.

We will now deal with the subsequent circuitry (the circuit functions following point "A" on the block diagram). The width-modulated pulses are fed to an active low-pass filter, or integrator. In this manner, a near-dc voltage is developed that is proportional in magnitude to the phase angle Θ. The active low-pass filter *also* performs as a summing amplifier, accommodating an adjustable dc voltage on one of its inputs. This adjustment is the power-factor command. It enables optimum performance to be attained for the load circumstances under which a particular motor operates.

The near dc output voltage from the low-pass filter, together with a line-synchronized ramp voltage, are applied to the input terminals of a voltage comparator. The output pulses from the voltage comparator are thereby position modulated, or timed, by the *amplitude* of this dc voltage (the *error voltage* in Fig. 6-2). This dc voltage represents phase-angle Θ, that is the power factor. Finally, by using the voltage-comparator output pulses (bottom waveform, Fig. 6-2) to trigger the triac, the effective voltage applied to the motor can be controlled much in the manner of ordinary light-dimmer circuits. More specifically, low power factor will cause *delayed* firing of the triac, thereby *reducing* motor voltage. As previously explained, this causes the motor to reduce its current and to increase its power factor. These, of course, are the requisites for efficient operation.

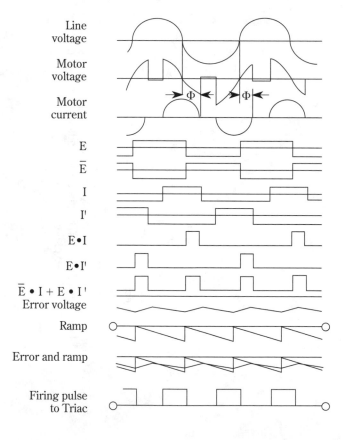

6-2 Waveform diagram pertaining to the power-factor controller. In response to power-factor angle, Φ, the pulse-train, $E.I + E.I'$ is developed. Then, the average value of these pulses, together with a line-synchronized voltage-ramp, controls the timing of triac gate-pulses. Thus, at low power-factor motor voltage is *lowered*.

Several side effects of such electronically controlled power factor can be mentioned. Because of cooler operation from lower current demand at light loads, motor longevity is enhanced. In industrial installations where many small motors, or fewer large ones, are used the penalty often imposed by power companies for low power factor can be avoided. The light-load speed reduction from reduced voltage will generally be of negligible amount, say on the order of two percent or less. Also, the concept is applicable to three-phase motors. The idea is to sense Θ in one phase, but to use the error voltage to control triac triggering in all three phases. Because of unique circuit requirements imposed by various polyphase arrangements with regard to delta, Y, and neutral connections, some ingenuity might be required for successful implementation.

The circuit shown in Fig. 6-3 is applicable to many single-phase fractional horsepower induction motors. Correspondences with the block diagram of Fig. 6-1 are as follows:

The voltage-squaring amplifiers are op amps IC1A and IC1B. The current-squaring amplifiers are op amps IC1C and IC1D. Diodes D1 and D2, together with their resistance networks perform the logic operations. IC2A is the summing amplifier and low-pass filter. The *power-factor command* adjustment is provided by potentiometer P1. The line-synchronized ramp generator is the 2N2907 pnp transistor. Op amp IC2B functions as the ramp and error-voltage comparator. A driver triac is used ahead of the actual motor-control triac. Two simple half-wave power supplies produce the required plus and minus 15 volts. They are regulated by zener diodes.

The IC2C circuitry is not referred to in the block diagram and has not been mentioned in the discussion of operating theory. This is an auxiliary circuit that delays triggering of the triacs for several seconds after the system is first turned on. It prevents an initial current surge tending to occur because the 15-volt supplies do not become immediately operational.

The op amps can be LM324s in which case IC2D of these quadruple units is not used. TR1, the motor-drive triac is a 25-ampere, 400-volt type, such as the 2N6161. This triac will need to be heatsinked.

The power factor controller is a government-owned invention, but licenses for commercial development are available at no charge. Further information can be obtained from the Technology Utilization Office of the Marshall Space Flight Center, Administration and Support Division, Marshall Space Flight Center, Alabama 35812.

The peculiar shape of the motor voltage waveform merits special mention. Because of the inductance of the motor's stator winding, the voltage applied to the stator and the resultant stator current are considerably out of phase. More specifically, the stator current lags the stator voltage. Once a triac is triggered into conduction, such conduction does not cease until the current through the triac passes through zero. This is true even though the voltage across the triac is reduced to zero (the *current* is the governing parameter). With lamps, or other resistive loads, both current and voltage pass through their zero values together, so it might *appear* that zero voltage is the cause of the triac going out of conduction. Despite the notches thereby produced in the motor current waveform, operation and control are essentially as good as would be the case with a more ideal voltage wave developed across the motor.

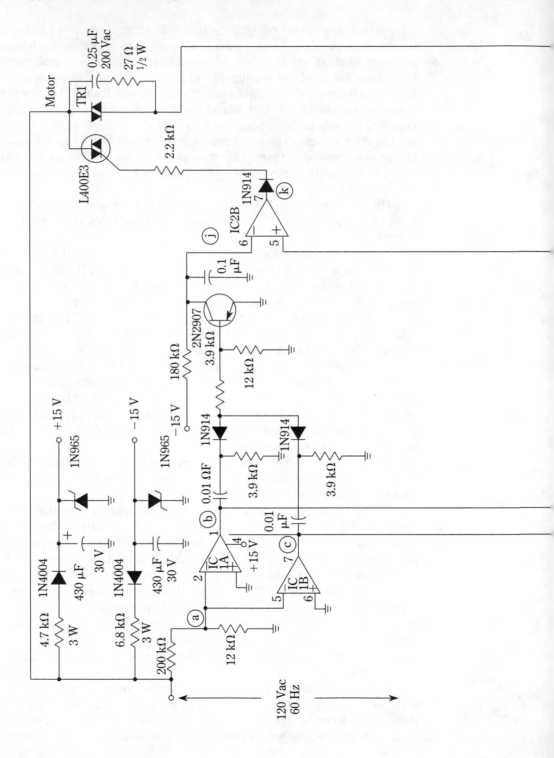

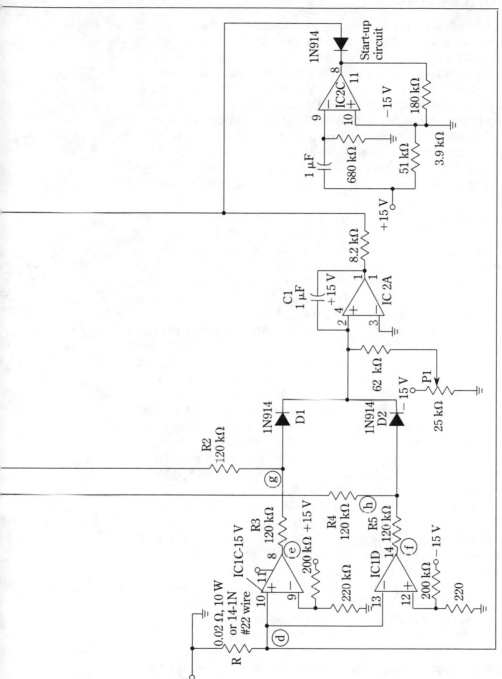

6-3 Circuit for the power-factor controller for fractional-horsepower induction motors. Triac TR1 can be typically a 400-volt, 25-ampere unit, much as the 2N6161. A heatsink is required for most applications.

Stepper motor controller

Integrated circuit technology and digital logic implementations can be associated with the stepping motor to bring about precise performance and positional control. For many applications, such a system is superior to conventional servo circuitry. For example, because there is no feedback, there are no instability problems. Also, no compromises are needed to avoid overshoot and oscillatory behavior in mechanical members. Best of all the control logic can be achieved via a single IC module. When so done, programming is accomplished with an ASCII keyboard. Sequences of high-level type commands can be stored internally in a program buffer and executed upon command. This is an excellent technique for prototype development. Figure 6-4 shows such a system with the CY500 stepper motor controller IC made by Cybernetic Micro Systems. The list of commands that can be stored and/or executed are shown in Fig. 6-5.

An interesting aspect of this control technique is that the stepping motor can be operated in the half-step mode. That is, it can be sequenced through twice as many discrete positions as would ordinarily obtain from the actual number of physical pole teeth. This, of course, doubles the resolution of the motor.

As an example of the "intelligence" that resides in such a programmable motor system, one could envisage a coil-winding machine in which it is desired to wind a number of layers of fine wire with a preset number of turns per layer. The process should take place at a known and stable rate—too slow is uneconomical; too fast might break the wire. Appropriate numerical values are selected for "N" and "R" and the programming can allow for repetition of the layer-winding activity until a signal from an external control line actuates the "Q" command to terminate the procedure. Thereafter, the procedure could be repeated, or there could be a return to the command mode. It is clear that the task would be accomplished in robot-like fashion with uniformity and precision. Although a similar overall routine could be instituted with a linear servo system, one would constantly have to be concerned with erratic torque, drifting limits, and transient conditions. Although "average" performance might be acceptable for reasonable time periods, the fine wire could easily be snapped by a low-level oscillatory condition even though it might be of short duration.

The pin configuration of the CY500 is shown in Fig. 6-6, and Fig. 6-7 is a logic diagram of this versatile IC. The parts-count reduction that is realized from incorporating the numerous functions within a single monolithic module is quite dramatic when compared to former control systems. Not only is this device programmable from the ASCII keyboard, but it is readily commanded from a microcomputer, from an eight-bit data bus, or from data stored in ROM, PROM, or EPROM. Moreover, it is feasible to incorporate a feedback path so that the stepping motor cannot be commanded to accelerate, reverse, or brake beyond its mechanical abilities. The timing and control signals involved in typical operating modes are shown in Fig. 6-8. The stepper control signals, Φ_1, Φ_2, Φ_3, and Φ_4 must be power boosted in order to actually drive the four-phase stepper motor. While this can be accomplished with discrete transistor stages, or with power ICs, special modules are available. An example is the Sprague ULN 2813A.

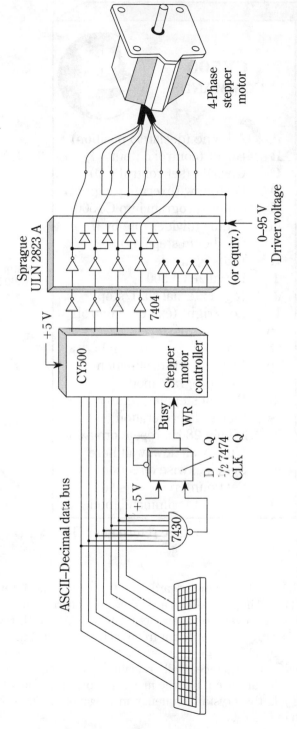

6-4 A prototype system utilizing the CY500 stepper-motor controller. An ASCII keyboard permits control of such motor-parameters as direction of rotation, stepping rate, number of steps, repetitive routines, etc. Cybernetics Micro Systems.

CY500 COMMANDS

A	Athome (declareØ position)
B	Bitset (control output = 1)
C	Clearbit (set control = Ø)
D	Do it now (execute program)
E	Enter (program into CY500)
Ff	Factor (divides rate by f)
G	Go (begin stepping)
H	Halfstep mode
I	Initialize CY500
J	Jog (external start/stop)
L	Left/right (ext. dir. control)
Nn	Number of steps n
O	Onestep (immediately)
Pp	Position p is destination
Q*	Quit program mode
Rr	Rate of stepping set to r
Ss	Slope of acceleration (±s)
T	Til pin 28 Hi, repeat program
U	Until 'wait' low, wait here
+	Set clockwise direction
−	Set counterclockwise dir.
Ø	Return to command mode

6-5 List of commands that can be stored or executed with the CY500 IC. Cybernetics Micro Systems.

* No carriage return after Q

The most straightforward way to actuate the internal clock circuitry of the CY500 is to simply connect a six megahertz series-resonant crystal to the appropriate pins. However, considerable flexibility exists with regard to the clock function. Crystals with series-resonant frequencies between 1 and 6 MHz can be used. In particular, it might be convenient to use a 3.58-MHz TV color burst crystal. In most instances, performance will remain satisfactory when these various crystal frequencies are used, but the motor stepping rate must then be scaled down by $f/6$, where f is the crystal frequency in megahertz. Three clock-circuitry options are shown in Fig. 6-9.

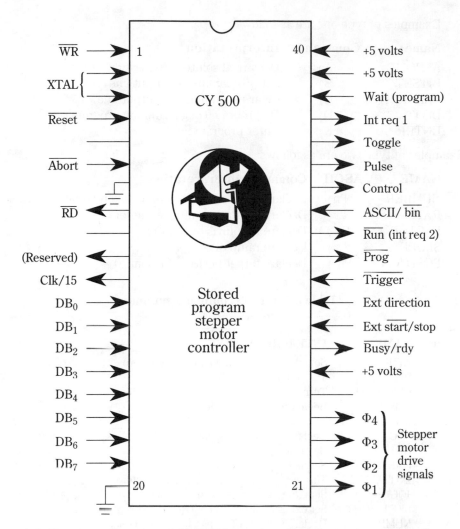

6-6 The pin configuration of the CY500 stored program stepper-motor controller. A single five-volt supply suffices to operate this device. Cybernetics Micro Systems.

Programming information is depicted in simplified form in the CY500 command summary (Table 6-1). More detailed explanations of such data appear in the description of commands in Table 6-2. In using these tables, every command entered consists of one of the following forms:

- Alphabetic ASCII character followed by the ')' (RETURN) key.
- Alphabetic ASCII character followed by blank, then ASCII decimal number parameter, then)' = 0DH.

Examples of type one are as follows:

Name	Command	Interpretation
ATHOME	A⏎	Declare absolute zero location
BITSET	B⏎	Set programmable output line
CLEARBIT	C⏎	Clear programmable output line
DO IT	D⏎	Do program (begin running program)
ENTER	E⏎	Enter program mode

Examples of type two are as follows:

NAME	ASCII	Command interpretation
NUMBER	N n⏎	Declare number of steps to be taken (relative)
RATE	R r⏎	Declare maximum rate parameter
FACTOR	F f⏎	Declare rate division factor
SLOPE	S s⏎	Declare ramp rate
POSITION	P p⏎	Declare target position (absolute)

Table 6-1. CY500 command summary.

ASCII CODE	NAME	INTERPRETATION
A	ATHOME	Set current location equal absolute zero
B	BITSET	Turn on programmable output line
C	CLEARBIT	Turn off programmable output line
D	DOITNOW	Begin program execution
E	ENTER	Enter program code
F	FACTOR	Declare rate divisor factor
G	GOSTEP	Begin stepping operation
H	HALFSTEP	Set halfstep mode of operation
I	INITIALIZE	Turn off step drive lines, initialize system
J	JOG	Set external start/stop control mode
L	LEFTRIGHT	Set external direction control mode
N	NUMBER	Declare number of steps to be taken (relative)
O	ONESTEP	Take one step immediately
Q*	QUIT*	Quit programming—re enter command mode. *Not followed by ' '
R	RATE	Set rate parameter
S	SLOPE	Set ramp rate for slew mode operation
T	Loop TIL	Loop 'TIL' external start/stop line hi
U	UNTIL	Program waits until signal line goes low
+	CW	Set clockwise direction
−	CCW	Set counterclockwise direction
0	COMMAND	Exit program execution mode, enter command execution mode

Table 6-2. Description of commands.

A) ATHOME   0100 0001   1 byte

The ATHOME instructions (sets the position) defines the 'Home' position. This position is absolute zero and is reference for all POSITION commands. The ATHOME command must not be immediately preceded by a change-of-direction command. Note also that the ATHOME command should not be used twice unless the CY500 is reset or initialized between commands.

B) BITSET   0100 0010   1 byte

This instruction causes the programmable output pin (#34) to go HIGH. This is a general-purpose output that may be used in any fashion.

C) CLEARBIT   0100 0011   1 byte

This instruction causes the programmable output pin (#34) to go LOW. The user can signal locations in a program sequence to the external world via B and C instructions.

D) DOITNOW   0100 0100   1 byte

This instruction causes the CY500 to begin executing the stored program. If no program has been entered, the controller will return to the command mode. If the program exists, the controller will begin execution of the first instruction in the program buffer. If the run (DOITNOW) command is encountered during program execution, it restarts the program (however, the initial parameters, and modes, may have changed) and may be used for looping or cyclic repetition of the program.

E) ENTER   0100 0101   1 byte

This instruction causes the CY500 to enter the program mode of operation. All commands following the ENTER command are entered into the program buffer in sequence.

F) FACTOR f   0100 0110   2 bytes
                            b7   b0

The FACTOR command causes the rate to be decreased by the factor (1/f). The factor, f, is a number from 1 to 255.

G) Go command   0100 0111   1 byte

The GO command causes the stepper motor to step as specified by the rate, direction, etc., commands entered prior to the GO command.

H) HALFSTEP command   0100 1000   1 byte

The HALFSTEP command causes the CY500 to enter the halfstep mode and remain in that mode until the device is reset or reinitialized.

I) INITIALIZE   0100 1001   1 byte

Table 6-2. Continued

The INITIALIZE command causes the CY500 to enter the command mode. *None of the distance or rate parameters are altered.* Any commands following 'I' will be executed with the parameters specified prior to 'I.' The INITIALIZE command, when encountered during program execution, halts the program execution and returns the system to the command mode. This command de-energizes the stepper motor coils.

J)   JOG   0100 1010   1 byte

The JOG command places the CY500 in the external start/stop control mode; i.e., the starting and stopping are controlled by external hardware instead of by software (via the GO command). *The 'J' command is the last command applied after the step rate, mode, and direction have been specified.* However, new commands may be entered when the XSS pin is low. It is NOT necessary to use the GO command. The application of a low voltage to the XSS pin causes the motor to begin stepping and continue stepping as long as pin #28 is low. Note that the 'J' command and the 'T' command are mutually exclusive. Either one or the other, but not both, may be used.

L)   LEFT/RIGHT pin enable   0100 1100   1 byte

This instruction places the CY500 in the external direction control mode. In this mode, the External Direction Pin (#29) is used to determine the direction, either clockwise (1) or counterclockwise (0). Software commands '+' and '−' are ignored in this mode. Any change of control signal applied following the output pulse will be used to determine the direction of the next step.

N n) NUMBER of steps   0100 1110   3 bytes
                                    a7   a0   LSbytes
                                    b7   b0   MSbytes

The NUMBER command is used to specify the number of steps to be taken in the 'Relative' mode of operation. The argument may be any number from 1 to 64K. Note that this parameter is stored as 2 bytes in the program buffer. An 'N' command *immediately* following the ATHOME command will take one step less than the specified number. (Note: current devices do not work correctly for n=multiple of 256. This will be corrected in the next batch of parts. Current devices do, however, treat all POSITION parameters correctly.)

O)   ONESTEP   0100 1111   1 byte

The ONESTEP command is used to take a single step. The steps are untimed, and the step rate is determined by the rate at which 'O' commands are received. In addition, each step may be triggered externally via the TRIGGER pin.

P p) POSITION   0101 0000   3 bytes
                                    a7   a0
                                    b7   b0

The POSITION command declares the 'Absolute' mode of operation. The argument is treated as the target position relative to position zero. The ATHOME command can be used to define position zero.

Q ⟩   QUIT (Programming)   0101 0001   1 byte

NOTE: The QUIT command is self-terminating, and should NOT be followed by the Linend '⟩'.

The QUIT command causes the CY500 to exit the 'Programming' mode of operation wherein instructions are stored in the program buffer in the order received, and to return to the 'Command' mode of operation in which instructions are executed as they are received. Note that the 'Q' command is *not* terminated with the Linend character, ØDH=⟩ , and such termination may result in incorrect operation. The QUIT command is also used after program completion prior to entering new parameters.

R r ⟩   RATE rate   0101 0010   2 bytes
                                            $b7$ $b0$

The RATE instruction sets the rate parameter that determines the step rate. The rate parameter, r, varies from 1 to 253 corresponding to step rates from 50 to 3350 steps / sec (assuming a rate factor of 1 and a 6 MHz crystal). The rate is nonlinear and can be computed from the rate equation. For crystals other than 6 MHz the step rate should be multiplied by f / 6 MHz, where f is the crystal frequency.

S s ⟩   SLOPE   0101 0011   2 bytes
                                            $a7$ $a0$

The SLOPE or slew mode of operation is used when high step rates are required and the initial load on the motor prevents instantaneous stepping at such rates. In such cases, the load is accelerated from rest to the maximum rate and then decelerated to a stop. The user specifies the distance of total travel (via 'N' instruction), the maximum rate (via 'R'), (FACTOR is forced to 1 automatically), and the ramp rate or change in rate parameter from step to step. When executing in the SLOPE mode, the CY500, starting from rest, increases the rate parameter by 's' with each step until the maximum (slew) rate is reached. The device computes the "re-entry" point at which it begins decelerating (with acceleration = −s) until it reaches the final position. NOTE: The user is responsible for ensuring that N>2 (maxrate).

T ⟩   loop TIL   0101 0100   1 byte

The 'T' command provides a 'Do . . . While . . .' capability to the CY500. This command tests pin 28 and, if low, it executes the DOITNOW command, i.e., it runs the program from the beginning (although using the latest rate, position, etc., parameter). If pin 28 is high, the next instruction is fetched and executed. Note that pin 28 is also used for external start / stop control when 'J' (JOG) is executed. 'T' and 'J' are therefore mutually exclusive and cannot be used together.

Table 6-2. Continued

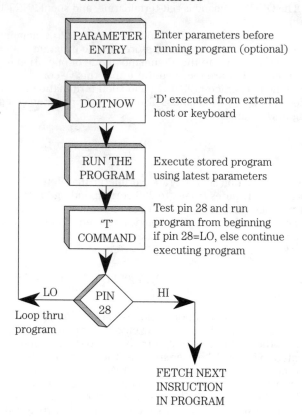

Example: Loop TIL command use.

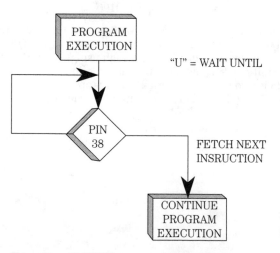

Example: wait UNTIL command use.

U)   wait-UNTIL   0101 0101   1 byte

The wait-UNTIL instruction is used to synchronize the program execution to an external event. When the 'U' instruction is executed it tests the WAIT pin. When the WAIT pin goes high, the next instruction is fetched from the program buffer and execution proceeds.

+)   CLOCKWISE command   0010 1011   1 byte

All steps following this command will be taken in a clockwise direction.

−)   CCW command   0010 1101   1 byte

All steps following this command will be taken in a counterclockwise direction.

0)   COMMAND mode   0100 0000   1 byte

The CY500 is placed in the command mode and the next command is executed as it is received.

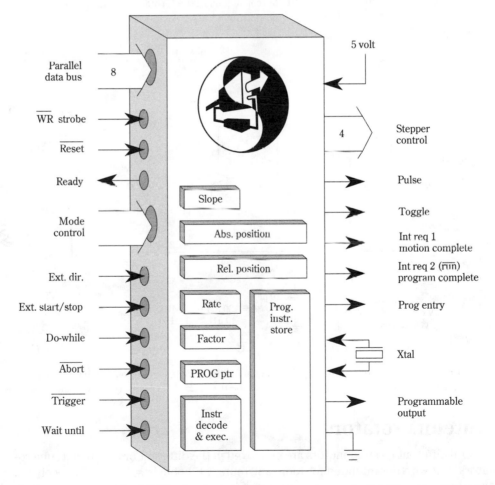

6-7 Logic diagram of the CY500 stored program stepper-motor controller. The stored program capability is 18 hi-level instructions. Cybernetics Micro Systems.

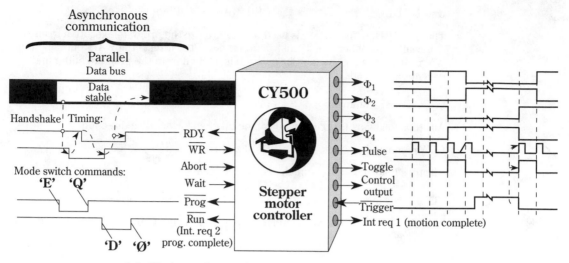

6-8 Timing and control signals in the CY500. <small>Cybernetics Micro Systems.</small>

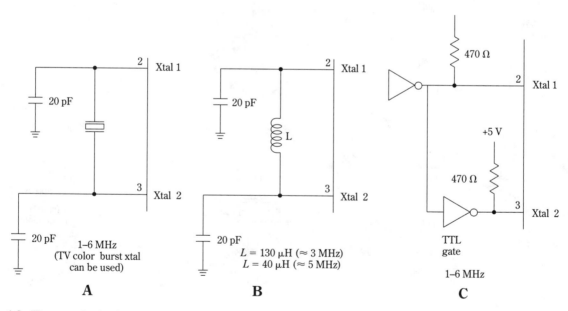

6-9 Three methods of generating the clock signal for the CY500. (A) Series-resonant crystal. Preferably 6 MHz, but frequencies between one and 6 MHz can be used. (B) LC circuit. (C) External. <small>Cybernetic Micro Systems.</small>

Antenna rotators

Several different types of motors are employed in the different brands of antenna rotators. But whether intended for radio amateurs, CB hobbyists, FM, or TV receivers, the basic control concept of automatic types is quite similar. Automatic operation infers that the system goes into a self-seeking operational mode that terminates when the antenna points in the direction selected by a dial, pointer, or other indicating device in the operator's control unit. Some of the performance attributes demanded of the motor in the drive unit are high-starting torque, reversibility, and extreme ruggedness and reliability. The dc series motor has excellent torque characteristics. To be reversible, such a motor must either have free access to both series field terminals, or a center-tapped field winding. Permanent-magnet dc motors also have good starting torque and can be readily reversed by reversal of the polarity of the applied voltage. What has been said with regard to the dc series motor also applies to the ac universal motor. Finally, biphase induction motors can be employed much as they are used for garage door actuation.

The most used electronic control technique is that of a bridge in which one arm comprises a variable resistance element mechanically coupled to the antenna. An opposing bridge arm contains a variable resistance mechanically linked to a directional indicating dial or device. The basic idea of this arrangement is that bridge unbalance tends to be corrected as the antenna is driven in the appropriate direction to null the bridge. At null, the power to the drive motor is cut off. Overall, such a system might be described as a position-seeking servo. All manners of electro-mechanical relays, contacts, SCRs and triacs are used in commercial units to perform the various switching functions. Figure 6-10 shows a block diagram of a typical automatic antenna rotator.

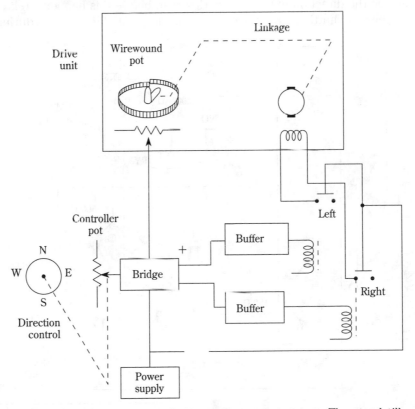

6-10 Block diagram of a typical antenna rotator system. The standstill torque capability of the drive motor is of paramount importance in such a positional servo-arrangement.

More insight into the actual implementation of electronic control circuitry is provided by the simplified circuit shown in Fig. 6-11. In this case, direction reversal is achieved by reversal of polarity in the voltage applied to a dc permanent magnet motor.

In Fig. 6-11 assume that the system is operational and is in equilibrium, i.e., the directional orientation of the antenna coincides with that indicated on the operator's azimuth dial. Under this condition, the bridge comprised of resistance elements R1 through R6 is balanced (R7 is in parallel with R6 and is a calibration adjustment). Notice that R2 is the azimuth dial and R5 is mechanically ganged to the motor and antenna shaft. You can see at once that the basic idea is to have R5 driven in such a direction to restore balance to the bridge when such balance is upset by adjustment of the azimuth dial. Another design objective is to have the motor turned off when the antenna points in the direction indicated by the azimuth dial.

Suppose that the azimuth dial is turned in such a direction that the resultant bridge unbalance causes transistor Q2 to be cut off because of the bridge voltage applied to its base-emitter circuit. Then transistor Q1 will also be biased in its nonconductive state. Relay RY2 will be deprived of solenoid current; in turn the motor will be polarized (through the normally connected relay contacts) to rotate in the clockwise direction. This goes on until the bridge approaches balance and Q2 and Q1 are biased into their conductive states. When this happens, relay RY2 is energized, tending to reverse the direction of the motor. However, before this happens, relay RY1 drops out, thereby shutting down the system. The drop out of RY1 occurs during the

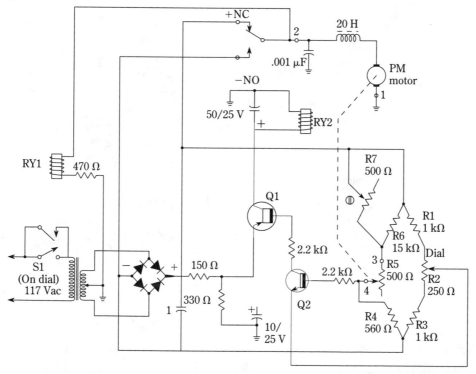

6-11 Electronic control circuit of a typical automatic antenna rotator. The bridge circuit has one arm mechanically-linked to the antenna; the opposing arm is controlled by the operator's azimuth-dial.

brief time required by RY2 to change its state. Once this happens, the whole system latches into a quazi-off state, with the motor being deprived of power. This is acceptable because the bridge is then balanced.

If the operator turns the azimuth dial, it not only changes the R2 arm of the bridge, but through appropriate mechanical linkage, it momentarily closes primary switch S1. Once this occurs, RY1 is energized, maintaining primary power until RY2 again undergoes a change of state as bridge balance is approached. Note that a momentary closure of S1 suffices to *initiate* operation; a momentary deprivation of voltage across RY1 when RY2 changes state, suffices to *halt* operation. The restorative action tending to null the bridge is essentially the same whether the bridge is initially unbalanced in one direction or the other. The basic position-servo techniques long used for rotating amateur and TV antennas are now being adapted for pointing the parabolic dish antennas used for satellite TV earth stations.

Speed and direction control for dc motors

There are many applications for dc motors where control of speed and direction is important, but it is satisfactory to allow speed regulation to be that produced by the natural characteristics of the motor. Thus, the control circuits shown in Figs. 6-12 and 6-13 do not make use of feedback. The circuit in Fig. 6-12 is intended for shunt motors, but will also accommodate permanent-magnet motors because reversal is accomplished through change of polarity of applied armature voltage. The circuit in Fig. 6-13 is intended for series or universal motors. These circuits are predicated upon classical principles of dc motor control, but do not use rheostats for reducing applied armature voltage. Instead, SCRs are used in a form of phase control. How this can be done and still deliver dc to the motors will be explained. Although components and semiconductor devices were chosen to comply with the needs of $\frac{1}{15}$-hp motors, the circuits can readily be "beefed up" to accommodate larger fractional-horsepower machines.

Consider first the circuit in Fig. 6-13. Note the connections of armature and series field relative to the bridge arrangement of SCRs. Current flow through the armature will always be in the *same* direction. However, the direction of current flow through the series field depends on whether conduction is provided by SCR1 and SCR4, or by SCR2 and SCR3. Thus, the direction of rotation is determined by the setting of switch S1 because it selects one or the other of these SCR pairs for triggering into conduction.

Speed control is achieved by timing the conduction through the selected pair of bridge SCRs. This is accomplished without any commutation problems. No filter capacitor is associated with the rectified dc from the diode bridge, D1, D2, D3, and D4. This is neither an oversight, nor a cost-cutting technique. Rather, the *pulsating* dc waveform from full-wave rectification is important in the operation of the circuit. It is because of the pulsating waveform that "natural" commutation takes place in the SCRs. If dc or nearly-pure dc were used, some means of periodically turning off the SCRs would have to be devised. There are circuits that operate in this way, but they often prove to be critical and unreliable. Failure to commutate SCRs can, at best, result in loss of control; at worst, serious damage can occur to the SCRs, or to the load.

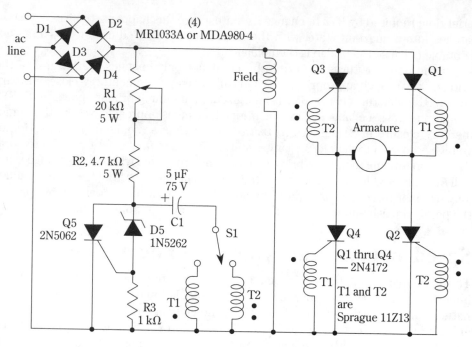

6-12 Control circuit for shunt and permanent-magnet motors. Change in direction of rotation is brought about by reversing armature current. <small>Motorola Semiconductor Products, Inc.</small>

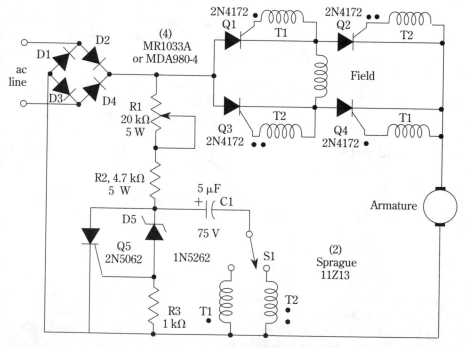

6-13 Control circuit for series and universal motors. Commutation of the SCR's ensues from the periodic zeroing of the full-wave rectified dc. <small>Motorola Semiconductor Products, Inc.</small>

It is interesting to trace out how the motor-control SCRs are triggered. This occurs as a consequence of capacitor C1 charging until it develops sufficient voltage to cause conduction through zener diode D5. This forward biases the gate of SCR Q5, causing the SCR to fire. The resultant pulse is then transferred either through pulse transformer T1 or T2, depending on the position of forward reverse switch S1 to the appropriate pair of motor-control SCRs (Q1 and Q4 or Q2 and Q3). SCR Q5 is then commutated (turned off) when its rectified anode-cathode voltage dips to zero. After commutation, a new cycle of events commences with the recharging of capacitor C1. Note that Q5 is commutated in exactly the same way as are the motor-control SCRs. Thus, any filtering at the output of bridge rectifier, D1 through D4, would be detrimental to the operation of the circuit.

Potentiometer R1 enables the speed of the motor to be controlled. The greater its resistance, the longer capacitor C1 requires to develop sufficient voltage to trigger SCR Q5. Inasmuch as this then reduces the conduction time of the motor-control SCRs, the motor experiences reduced voltage and runs slower. The converse situation applies when the resistance of R1 is lowered; the motor control SCRs then conduct earlier in the ac cycle and conduct for a longer time. This increases the effective voltage impressed on the motor and it therefore runs faster.

The operation of the shunt and permanent-magnet motor control circuit of Fig. 6-12 is essentially similar. When a shunt motor is used, no attempt is made to control the voltage applied to the shunt field; motor control is achieved via control of armature voltage. When a PM motor is used, there is no field winding. But control remains by means of the voltage applied to the armature terminals. Reversal of rotation in this circuit is brought about by reversing the direction of current flow through the armature. This contrasts to the reversal technique used for series motors in Fig. 6-13 where reversal is attained by changing the direction of current through the series field.

Motor speed control by resistance modulation

For many years, a popular speed control for heavy duty series motors was a manually-actuated selector of resistance elements. For example, the motorman's controller used in street cars was just such a device. Despite its brute-force approach and maintenance problems, such a system actually works quite well and now merits reconsideration in applications to electric automobiles. A refinement of such a speed control method is shown in Fig. 6-14. The basic simplicity of resistance selection is retained, but speed control is achieved with much less power dissipation. This is of paramount importance in a vehicle that carries its own battery energy source.

In Fig. 6-14 the SCR can be pulsed on and off at a controllable duty cycle. When in its conductive state, the electrical losses of the SCR are relatively low—much lower than would be dissipated in "ohmic" resistance required to produce the same motor performance. The greater the duty cycle of the SCR, the more it effectively shunts current around a portion of motor resistance, such as R3. If the SCR duty cycle is adjusted for 180°, it behaves as a closed switch, essentially shorting whatever resistance it is connected across. These facts can be deployed in a clever way to accomplish a large range of smooth speed control.

Suppose, for example, the motor is moderately loaded and the SCR is continually in its *off* state. Because of the overall resistance in the motor circuit, you might assume the motor to be at standstill. If then, the SCR is triggered to provide fixed-

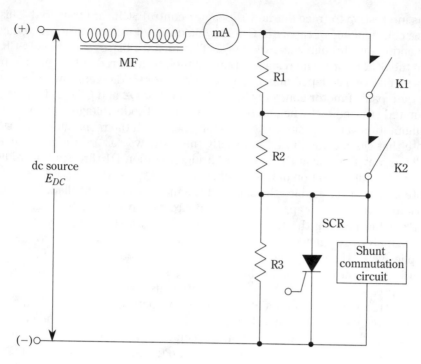

6-14 Series-motor speed control via resistance modulation. The SCR functions as a chopper to periodically short a portion of resistance in the motor circuit. Variation of the chopping rate yields a considerable range of smooth speed control.

duration closures at a low repetition rate, the average value of motor current will be increased because of the periodic shorting out of resistance *R3*. The motor will start and turn slowly. Increasing the trigger repetition of switch K2 will provide an additional range of speeds enabling faster motor operation. Finally, closure of switch K1 allows the third, and highest speed range to be obtained. Various links between the shorting switches and the SCR trigger source can be implemented for convenience and smoothness of control. The scheme is reminiscent of the formerly used series-motor control box in street cars, but the SCR pulsing technique yields better efficiency than in the older schemes that relied completely on dissipation in resistance elements. Of course, some kind of commutating scheme must be provided in order to turn off the SCR if a "pure" dc source is used. Self commutation can be realized with unfiltered half-wave, or possibly full-wave source current from a rectifier.

If the motor is not too large, it would be easier to implement this motor control concept with bipolar power transistors or with power MOSFETs. Then there would be no commutation problem. Such control devices could be paralleled to accommodate the current demands of fractional-horsepower motors. In particular, power MOSFETs lend themselves well to paralleling; no ballast resistances are needed to bring about equitable current divisions between paralleled devices, and unlike bipolar types, there is no "built-in" tendency towards thermal runaway.

In this control scheme, a somewhat different procedure might be necessary for heavy motor loads. Under such conditions, it would be more appropriate to start with switches K1 and/or K2 already closed. The basic idea would then be to provide *high* initial starting current for the motor. As the motor comes up to speed, either or both of these switches could be opened. The exact control procedure would depend on the desired acceleration and ultimate speed of the motor. Also, somewhat different procedures might prove more appropriate for shunt or permanent magnet motors than for series motors. The series inductance, MF, is not always needed, but is helpful in preventing excessively large surge currents during startup, and with heavy motor loads.

The dc regulated power supply as a motor controller

For the experimenter and innovator, the regulated dc power supply is an excellent basic system for controlling motor performance. This is particularly true if the supply has one or two remote-sense terminals—as in many bench-type supplies. Because of the current demands of motors during startup and occurring as a consequence of overloading, the ease with which this can be accomplished depends greatly on the application. In some instances, the power supply will have to be slowly brought up to normal operation by means of a *variac transformer* in the ac power line. If the motor is not subjected to excessive overloads, the electronic current-limit provision on many of these power supplies will provide sufficient protection. In certain cases, a small resistance must be placed in series with the motor. It might be necessary to use a series-pass or switching transistor of greater power capability than the original device. Remember that peak current demands of motors might last for many seconds as a nearly-stalled motor recovers speed. Summarizing, regulated power supplies have the brains to control motors—if any problem is encountered, it is likely to involve muscle and stamina. On the other hand, the small motors used in many electronics applications will generally be easy enough to deal with.

The control techniques to be described can be implemented with series-pass or with switcher-type supplies. In the first example, shown in Fig. 6-15, a series or PM motor is caused to have a constant torque. This stems from the fact that the connection scheme makes the regulating supply behave as a constant-current source, and motor torque is a function of its armature (and series-field) current. Indeed, examination of this scheme reveals that it is essentially identical to the way in which these supplies are ordinarily used to deliver constant current into a resistive or electronic load.

You can control and stabilize any type of dc motor with the arrangement shown in Fig. 6-16. Although perhaps not immediately obvious, this scheme simulates the conventional use of the supply when it acts as a voltage regulator. The basic idea is that the sensing leads do not "care" whether the error signal derives from a resistive or electronic load, or from the motor tachometer. In either case, output voltage will be stabilized. It fortunately happens that a constant armature voltage also tends to develop constant motor speed.

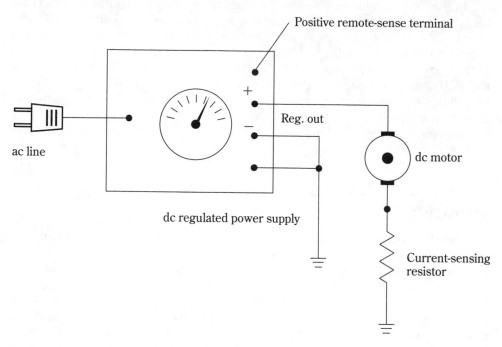

6-15 Using a dc regulated power supply to produce constant torque in a dc motor. The regulated supply is caused to operate in its constant-current mode. In the interest of efficiency, the current-sensing resistor should be as low as possible.

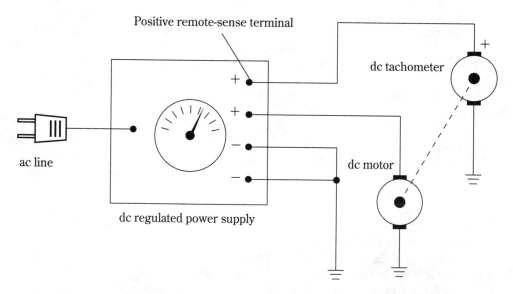

6-16 Using a dc regulated power supply to produce constant speed in a dc motor. The dc tachometer is a small PM generator mechanically coupled to the motor.

Precise control of motor speed with programmable divider

The arrangement illustrated in Fig. 6-17 enables precise control of motor speed without resorting to feedback or servo techniques. On the one hand, the reference oscillator and the programmable divider provide precise frequencies for the two-phase synchronous motor. On the other hand, a synchronous motor operates *only* at the speed determined by its construction and the frequency of the applied voltage. Applying heavier mechanical loads does not slow such a motor until a critical loading is reached, whereupon it comes to a standstill.

A two-phase motor requires voltages displaced by 90° for its two identical stator windings. This is commonly obtained, or approximated, with a capacitor connected in one winding. However, when different frequencies are impressed upon the motor, as in this application, it becomes necessary to switch in different capacitors to retain the quadrature phase relationship. To avoid such an inconvenience, a digital phase splitting method is used in this circuit. The two flip-flops form a two-bit ring that delivers quadrature-related square waves *regardless* of the frequency. At the same time, this ring performs an additional frequency division by a factor of four. This is easily taken into account in the programming of the frequency-divider proper.

The two quadrature-displaced square waves are processed in low-pass filters and power amplifiers, and fed to the motor. The low-pass filters modify the square

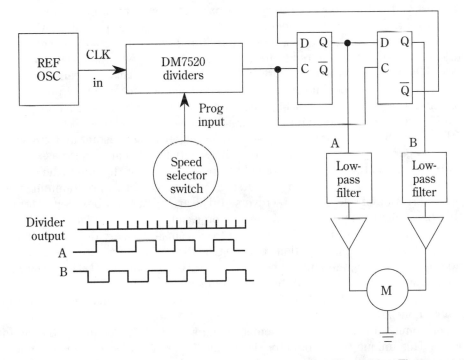

6-17 Speed control of a synchronous motor with a programmable divider. The two stator windings of the two-phase motor are quadraturely driven regardless of frequency.

waves, producing a smoother waveform for the motor. This is often desirable, for the harmonic content of square waves serves to aggravate eddy current and hysteresis losses rather than develop torque. If a wide speed range is implemented, it might prove beneficial to switch the cut-off frequency of the filters so that a near sine wave is applied to the motor regardless of selected speed.

There is more than meets the eye in this digital-control scheme; for instance, reversal of motor rotation might be brought about by switching logic-level voltages, such as interchanging connections at points A and B. A crystal-controlled reference oscillator will virtually transfer its high stability to the rotational speed of the motor. Operation is generally superior to that of conventional servo systems, because there is no feedback instability, "hunting," or dead zones. A possible improvement would be to increase motor voltage at high speeds and lower it at low speeds. This would enable the ac current in the phase windings to be within acceptable limits if wide-range speed control is attempted.

Constant-current motor drive

Manufacturers of regulated power supplies have been slow to adapt and specify their general product lines for motor-drive applications. Because of their apprehension of inductive loads and unpredictable current demands, substantial marketing opportunities surely have been lost. From the user's vantage point, many application requirements could be conveniently satisfied with appropriate associations between dc motors and suitable power supplies. For example, a supply operating in the constant-current mode is likely to meet the needs of a motor drive application where inordinately high starting torque is not required and where the motor is subject to overloads and stalls. If operated from an ordinary unregulated supply, damage could occur to both the motor and the supply during such times. This is because a nonrotating dc motor generates no counter EMF and becomes, for practical purposes, a short circuit. Fuses and circuit breakers often turn out to be unsatisfactory or inconvenient protection techniques.

Figure 6-18 shows a motor-drive system for maintaining constant motor current under all operating conditions. The circuit is essentially a linear, series-pass regulator with bipolar output. By means of voltage programming at the input, the desired motor current can be selected. By reversing the polarity of this programming voltage, the direction of rotation can be reversed. The arrangement is best suited for use with a permanent-magnet motor. A shunt motor can be accommodated if arrangement is made for supplying its field current from a small auxiliary dc source. If a series motor is used, the direction-reversing feature will be sacrificed insofar as concerns voltage programming. Of course, reversal of rotation can still be had via appropriate switches to transpose the connections of either the series field or the armature. As shown, a 48-volt motor can be controlled with a maximum constant current of one ampere.

The dominant feature of the schematic diagram is the ICL8063 integrated circuit module. This unique semiconductor device facilitates the drive of complementary symmetry power transistors from very ordinary op amps, such as the 741, as shown. By its use, the requirements of balanced output, protection against short circuits and against overdrive are automatically assured. Even though the input op amp operates

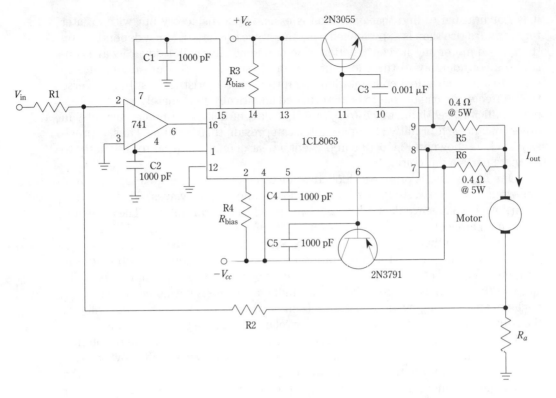

6-18 Constant-current motor drive. The complementary-symmetry drive IC provides current-regulated bipolar output for the motor. Selection of R1, R2, and R_a is discussed in the text. _{Intersil.}

from a ±12-volt supply, the output power transistors operating from a ±30-volt supply will be driven to full output. As previously mentioned, full output for this particular motor-control arrangement is one ampere. This one ampere output current will prevail even with the motor in a stalled condition. Another interesting aspect of the circuit is that the 741 op amp does not need an auxiliary power supply. This is because the LM 8063 module has an internal ±13-volt regulated supply (these voltages being available from pins 1 and 15).

The one ampere constant motor current can be realized by making Ra one ohm and R1 and R2 each 10 kΩ. Then, ±1 volt for V_{IN} to the op amp will produce ±1 ampere in the motor. The operational mode of this control technique is such that the motor current is governed by V_{IN}. However, once V_{IN} is selected, the motor current remains fixed regardless of motor speed, loading, or temperature.

The use of square waves to operate motors

Because of the increasing popularity of solid-state inverters, the use of square waves to operate motors has become commonplace. A wide variation of results has attended such practice. At best, the substitution of such waveforms for the intended sinusoidal excitation has caused no problems. At worst, it has been found that the motor would not start or operate properly. Between these two extremes, it is not unusual to find that operation is essentially satisfactory but with greater temperature rise than is experienced with sine-wave operation. Much depends upon the type of motor and its construction. The harmonic energy in a square wave contributes considerably to the eddy current and hysteresis losses in motors. It also tends to adversely affect commutation and torque characteristics. Just how adverse these effects are, usually has to be determined empirically. Fortunately, some motors have sufficient self-inductance to oppose the flow of heavy harmonic currents. In other instances, a primitive low-pass filter can provide this function. Thus, a provision such as shown in Fig. 6-19 can present a near enough sinusoid to the motor to alleviate certain difficulties.

One factor that is often overlooked when attempting square-wave operation of motors is the *equivalency* between sinusoidal and square waves. That is, what square-wave voltage would substitute for the rated sinusoidal voltage? The answer is not immediately obvious and if you resort to "common sense" you can easily be lead astray. In solving this dilemma, it is best to assume that the fundamental component of the square wave is useful in operating the motor. As previously mentioned, the harmonics either "come along for the ride" or they produce detrimental effects. To simplify the situation, suppose the motor's inductance keeps harmonic currents at a negligible level. The question that must then be answered is what the RMS value of the square wave's fundamental frequency is relative to that of the square wave itself.

Figure 6-20 shows the somewhat surprising fact that the peak value of the first harmonic in a square waveform *exceeds* that of the square wave itself. How can this be? The explanation is that the *overall* contributions of the fundamental and the many higher harmonics is such that the flat top of the square wave is developed. For example, if you considered the third harmonic, you would find that it *subtracts* from the peak of the fundamental. This is visually evident in Fig. 6-21. Again, the net re-

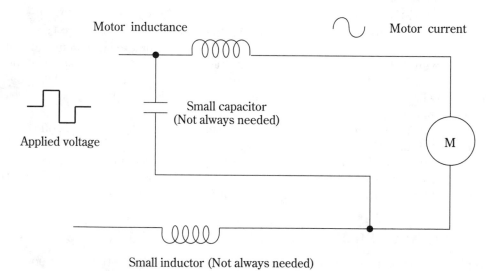

6-19 Filtering of motor current by self-inductance of motor and/or additional filtering. Actual motor current tends to be a rounded trapazoid. Additional filtering might be needed, but small filter components generally suffice because a high-quality sine wave is not necessary.

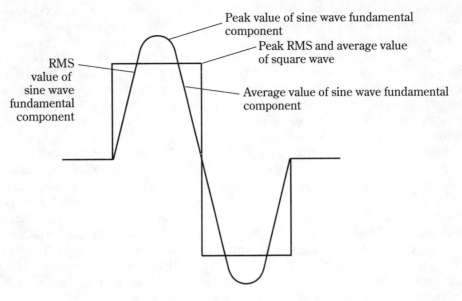

6-20 A closer look at relationships between a square wave and its fundamental. The peak of the fundamental sine wave is actually greater than the peak of the square wave, yet the RMS and average values are *lower* than those of the square wave.

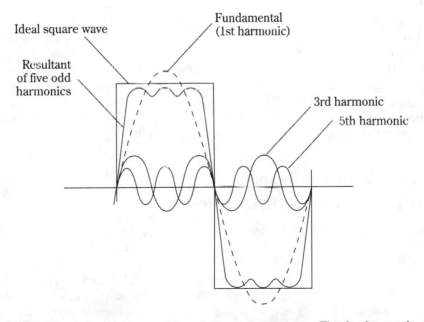

6-21 The harmonic composition of a near square wave. The fundamental (1st harmonic) produces motor torque. The higher harmonics generate eddy current and hysteresis losses and tend to interfere with smooth torque.

sult of *many* harmonics is the square wave, even though the peak of the fundamental actually exceeds the amplitude of the square wave itself.

It can be shown that the peak value of the fundamental sine wave exceeds the top of the square wave by the fraction $4/\pi$, or 1.27. In other words, the peak value of the fundamental is 27 percent greater than the flat-top, or peak value, of the square wave. The important thing to note, however, is that despite this relationship between peak values, the RMS value of the fundamental sine wave is still *below* that of the square wave. This too can be ascertained from Fig. 6-20. A *larger* amplitude square wave is needed in order to yield a fundamental sine wave whose RMS value matches the rated RMS sine-wave voltage of the motor. Thus, a 115-volt motor will require a square wave with a *greater* value than 115 volts. But, how much greater?

Assuming only that a particular motor has sufficient inductive reactance to cause motor current to be "reasonably" sinusoidal, you can calculate the required square-wave voltage that must be applied to the motor to simulate its rated sine-wave performance. We are not dealing with the low distortions expected in stereo equipment. Rather, distortions of 20 or 30 percent will enable entirely satisfactory motor operation. Referring again to Fig. 6-20, start with the fact that the peak value of the fundamental sine wave is 1.27 times the square-wave amplitude. The RMS value of this sine wave will then be 0.707×1.27, or 0.898 times the amplitude of the square wave. Thus, the sine-wave RMS value is actually *less* than the square-wave amplitude. (Square waves, RMS, average, and peak values are identical.)

The previous calculation shows that the requisite square wave must be 1/0.897, or 1.11 times the square-wave amplitude. For example, a motor rated for 115 volts from a sine-wave source will require 115×1.11, or 127.7 volts from a square-wave source, such as a solid-state inverter.

If there is not sufficient self-inductance, or external inductance associated with the motor, the above reasoning might not prove valid. Indeed, the first several harmonics of the squarewave might then contribute enough hysteresis and eddy-current heating, and might interfere sufficiently with the torque of the motor so that an actual *reduction* in square-wave voltage from the motor-rated voltage might prove necessary. Under such conditions, the optimum performance will not likely be forthcoming. A rudimentary low-pass filter in conjunction with a square-wave voltage of 110 percent rated motor voltage will yield good results in most instances.

ac operating voltage from the Darlington inverter

At an earlier developmental stage, the Darlington power transistor was not looked upon favorably as a power device for actuating and controlling motors. Inasmuch as the Darlington transistor does not *saturate*, it was argued that relatively high power dissipation would lead to low operating efficiency. And, the early Darlington transistors displayed obviously poor characteristics for motor work—they had inadequate power-handling capability and were easily damaged by current surges and voltage transients. Several things have happened to change this picture. First and foremost, modern Darlington transistors have become respectable power devices. As such, they are electrically rugged to the extent that they merit serious consideration as substitutes for SCRs in many applications. This is readily seen from the availability of Darlington transistors capable of withstanding 400 volts and with 20-ampere current capability. Collector dissipation, because of nonsaturation, is of negligible consequence at high operating voltages, say above 100 volts.

Interestingly, it is one of the Darlington's alleged "drawbacks"—inability to saturate—that now compels its use in motor-drive systems. Its inability to saturate is largely responsible for its efficient chopping operation in the 20-kHz region. This enables the highly efficient and flexible pulse-width modulation technique to be implemented at frequencies where SCRs are not likely to be at their best. This is especially so considering the freedom gained from *commutation* problems when substituting the Darlington transistor for the SCR.

Figure 6-22A shows an array of Darlington transistors arranged to provide ac power to a three-phase motor. The input waveforms necessary for such direct production of motor drive power are shown in Fig. 6-22B. The best use of the Darlington inverter results when the bases are driven with high-repetition rate width-modulated pulses, as shown in Fig. 6-22C. The use of this waveform enables automatic variation of *both*, the frequency of the reconstituted sine wave and its amplitude (an induction motor's voltage requirement is inversely proportional to frequency).

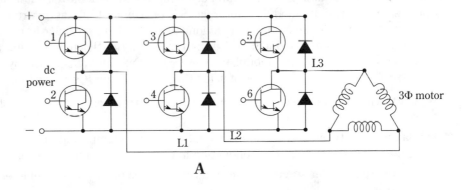

A

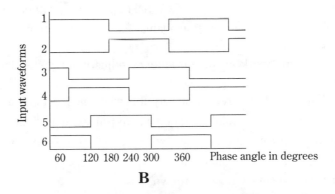

B

6-22 The use of a Darlington inverter to supply ac motor power. (A) Array of Darlington switching transistors for inverting dc to three-phase ac. (B) Base input waveforms for direct development of desired drive frequency. (C) Fabricated sine wave when pulse width modulation is used. The pulses have a repetition rate much greater than the simulated sine wave.

6-22 Continued

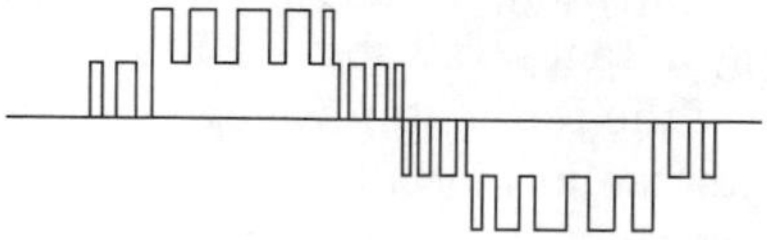

C

Considerable simplicity in inverter design is afforded by the modern monolithic Darlington device. Not only are both transistors integrally packaged, but so are the emitter-base resistors and the free-wheeling diode.

Pulse-width modulation in a package

One of the important methods of controlling the performance of motors is by inverters that are pulse-width modulated. Pulse-width modulation can be brought about by circuitry made up of discrete devices, with appropriate use of general-purpose ICs, or with a *dedicated* IC designed expressly for the function. The latter approach is probably the best and affords a neat and economical way to achieve a vital circuit function with a single IC module. Motorola, Silicon General, Texas Instruments, Presley, and National Semiconductor are among those marketing similar IC modules for achieving pulse-width modulation. No matter how large the motor being controlled, such an IC pulse-width modulator can be used as the logic for the system. Some of the features and advantages of using these specialized ICs are as follows:

- Worthwhile savings in cost, production time, and troubleshooting. Ease of duplication.
- Maximum duty cycle is about 45 percent, therefore eliminating the possibility of destructive simultaneous conduction in push-pull or bridge inverters.
- Dual-output driver stages enable the use of either push-pull or single-ended inverters.
- Current-limit amplifier provides external component protection.
- On-chip thermal-limit protection against excessive junction temperature and output current.
- External RC network enables frequency adjustment generally to well over 100 kHz.
- Five-volt, 50-mA linear regulator output is available for auxiliary purposes.
- Symmetry control by an injected dc voltage (on some models).
- Oscillator synchronization by injected frequency (on some models).
- Dead-time adjust (on some models).
- Inhibit terminal (electronic shutdown).

If for no other reason, you can expect superior performance from these pulse-width modulator ICs because of the elimination of numerous connecting leads which necessarily attend the use of discrete devices, separate operational amplifiers, and gates. Often well-founded fears of unstable and erratic circuit operation need no longer be entertained. With the advent of these single-chip modules, pulse-width modulation is nearly as easily realized as straight amplification. Moreover, in applications where wide-range speed control of induction motors is required, it is feasible to simultaneously obtain both frequency and amplitude control of the voltage applied to the motor (induction motors require higher applied voltage as the frequency is increased).

The LM3524 regulating pulse-width modulator IC

The LM3524 regulating pulse-width modulator IC is typical of similar units made by several semiconductor firms. Figure 6-23 shows the block diagram of the internal circuitry of the LM3524. Pulse-width modulation is accomplished by applying a sawtooth wave and a dc voltage to the inputs of a comparator op amp. The level of the dc voltage varies with that of the error signal—a sampled portion of the regulated output voltage. In response to this variation, the pulses appearing at the output of the comparator are duration modulated. Multiple-input NOR gates then drive a pair of output transistors. These transistors have free emitter and collector terminals and can be used for a wide variety of push-pull, bridge, and single-ended inverters and switching stages.

Unfortunately, there has been a lack of standardization in these pulse-width modulator ICs. Figure 6-24 shows the dual-in-line packages of two manufacturers. While basically similar in ultimate function, these ICs are actually not identical. Yet, both of them could provide very similar electrical performance when appropriately connected in many applications.

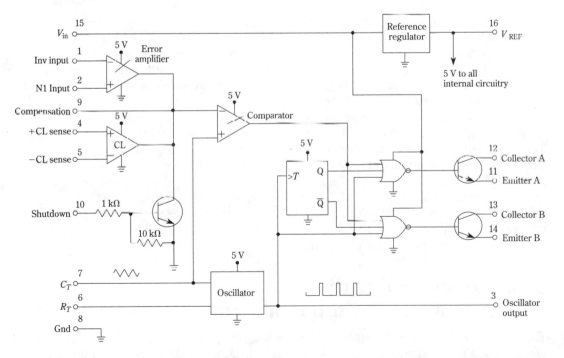

6-23 Block diagram of the LM3524 regulating pulse-width modulator. The undedicated output-transistors provide optional drive for single-ended, push-pull, bridge, or dc series-pass transistors. National Semiconductors Corp.

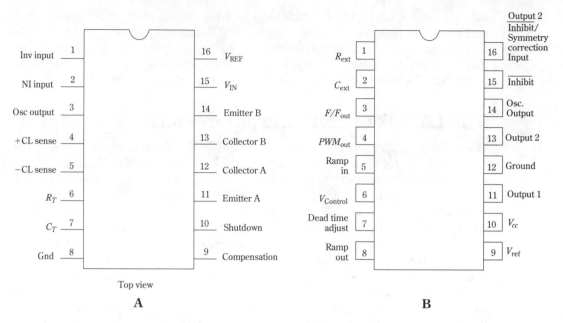

6-24 Pin connections of similar pulse-width modulators made by two manufacturers. Example of non-standardization. The two ICs provide nearly the same functions. (A) National Semiconductors LM3524. (B) Motorola MC3420.

In the National LM3524, as well as in other firm's products, the switching frequency is determined by an external capacitor and resistor. For the LM3524, Fig. 6-25 shows the oscillator period as a function of different combinations of the external capacitor and resistor. By plotting the chart in this manner, a family of nearly straight lines result. Frequency is the reciprocal of period. However, the frequency developed by a driven push-pull inverter will be half the oscillator frequency.

Of importance in some inverter situations, the dead time is greater at a given frequency if a large capacitor and small resistor are used to set the oscillator time constant. This is shown in Fig. 6-26. The principal use of dead time is to prevent simultaneous conduction in the push-pull or bridge transistors of a driven inverter. Much more than the needed dead time can degrade efficiency and interfere with regulation, however.

A typical inverter using the LM3524 is shown in Fig. 6-27. Circuits of this kind are very useful for providing selected speed options for induction and synchronous motors. Minimal filtering can be implemented, when satisfactory motor performance is not obtained from the output waveform.

dc motor control with the use of a pulse-width modulator IC

The possibility of controlling dc motors with the use of regulated power supplies was discussed in a previous chapter. There, it was mentioned that either a series-pass (linear) or switching-type regulator could be used. Moreover, it is quite easy to build a high-performance switching supply by using the pulse-width modulator ICs now avail-

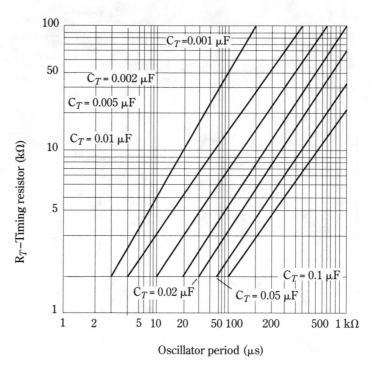

6-25 Oscillator period of the LM3524 as a function of the external capacitor resistor. This graphical technique provides a family of nearly straight lines. Oscillator frequency is the reciprocal of the period. National Semiconductor Corp.

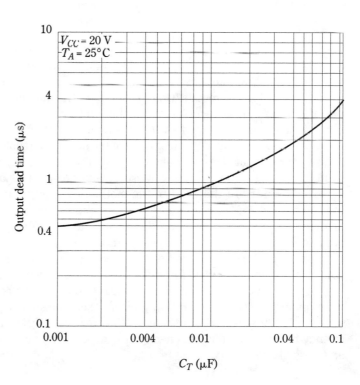

6-26 Deadtime developed by the LM3524 as a function of the external timing capacitor. For a given switching rate, deadtime is greater with larger capacitors and smaller timing resistors. National Semiconductor Corp.

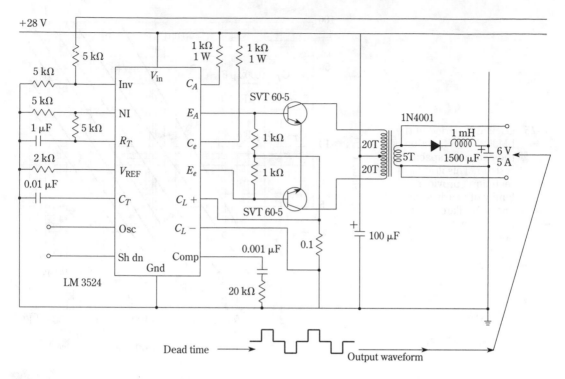

6-27 Push-pull inverter using the LM3524 regulating pulse width modulator. Deadtime in waveform prevents destructive simultaneous conduction in the push-pull output transistors.

able. Thus, unlike the situation prevailing in past years, the high-efficiency inherent in the switching technique can be readily realized without acquiring a manufactured supply. Only a few components need be associated with such a dedicated IC in order to obtain satisfactory results. Gone are most of the difficulties with instability, erratic performance, and catastrophic destruction of semiconductor devices that usually attended the breadboarding of switching supplies via the discrete-device approach.

Figure 6-28 is an example of the use of the switching-regulator approach to dc motor control. Notice that the switching regulator proper is comprised of the LM3524 pulse-width regulating IC, the switching transistor, and a handful of passive components.

The circuitry to the right of the motor provides the function of an *electronic tachometer*. This is only one of several techniques that can be used to obtain the dc error signal to apply to the LM3524. A small dc generator of the permanent-magnet type could be mechanically coupled to the motor. An ac tachometer, together with a rectifier and filter, could also be employed. In such an application, too great a time constant on the part of the filter could produce sluggish response, as well as overall loop instability. And optical methods with encoder disks could be used in a somewhat analogous way to the rotating-magnet scheme shown.

The LM2907 is also a dedicated IC, providing in this case, the function of frequency to dc converter. This neatly circumvents the problems of sluggish response and possible instability.

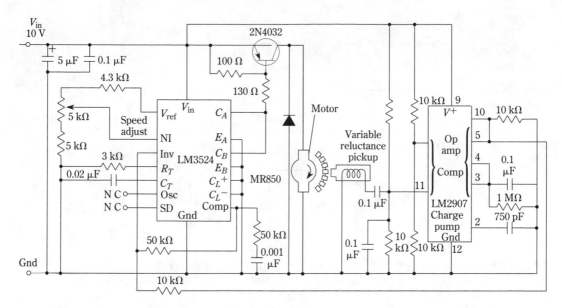

6-28   Use of the LM3524 pulse-width modulator IC to control a dc series motor. The circuitry to the right of the motor is, in essence, an electronic tachometer, which feeds a dc error signal to the LM3524.

The interesting aspect of such control systems as this one is that the "textbook" characteristics of the motor no longer apply. Rather, the speed, torque, and start-up behavior are very much governed by the electronic control logic. In some instances, for example, quite similar performance can be forthcoming from either a series, or a shunt motor! Without the electronic control, such motors would, of course, exhibit widely different characteristics.

Stepped-waveform inverter suitable for operation of motors

Although motors tend to operate quite well with square waves, another nonsinusoidal waveform can be produced that will more closely approach the efficiency and temperature rise that motors are rated for when driven by sine-wave power. Other improvements that can be brought about with this waveform are decreased noise and smoother torque. The waveform referred to is the stepped wave illustrated in Fig. 6-29. Interestingly, this quasi-square wave has the same peak and RMS values as the enclosed sine-wave depicted in dashed lines. Because of this, we find statements in the technical literature that say a motor driven by such a stepped wave essentially "sees" a sine wave. This is not quite true, but in practice it is found to be very nearly so, especially because the inductance of the motor exerts filtering action on the high-frequency harmonics of the wave.

Note, however, that what has been stated applies only to the special stepped wave in which the horizontal sections are 90 electrical degrees in duration. This structure is evident in Fig. 6-29; if this one-cycle wave were continued for many cycles, the 90° durations would be clearly visible.

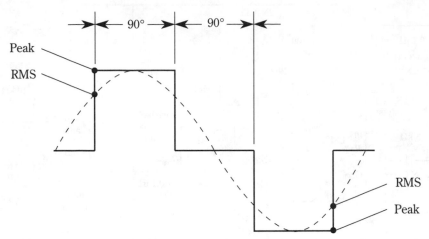

6-29 Unique stepped wave with peak and rms values identical to a sine wave. A motor impressed with such a waveform should behave very much as with the enclosed sine wave.

The schematic diagram of an inverter for producing the described stepped waveform from a 12-volt dc source is shown in Fig. 6-30. Although the power output is in the 40–50 watt region, circuit modification for higher power levels is relatively easy. Larger power MOSFETS are available, and paralling is easy with these devices. A replacement output transformer for T1 would also be required. However, larger power MOSFETS would be just as easy to drive as the prescribed ones. The salient feature of this circuit is the "brainy" portion that creates the desired stepped waveform. This comprises, for the most part, IC2, the decade counter, IC1, the quad op-amp, and Q7, used as a polarity inverter.

As shown, the stepped-wave inverter outputs a frequency of 75 Hz. This frequency was selected in order to provide a safety margin against core saturation of output-transformer T1. However, 60-Hz operation can be obtained by increasing the value of either R27 or C1. To a first approximation the factor of increase for either of these timing-circuit components will be 1.25. Probably a little experimentation will be needed because other factors bear a secondary effect on the oscillation frequency of IC1-a. Protection against magnetic saturation of T1 can be realized by using a larger transformer than would be dictated by only its current-carrying capacity.

The waveform diagram of Fig. 6-31 shows how the pulses from decade-counter IC2 are combined to produce the desired stepped waveform at the output of the inverter. The parts list is depicted in Table 6-3. This is an excellent project for the experimenter

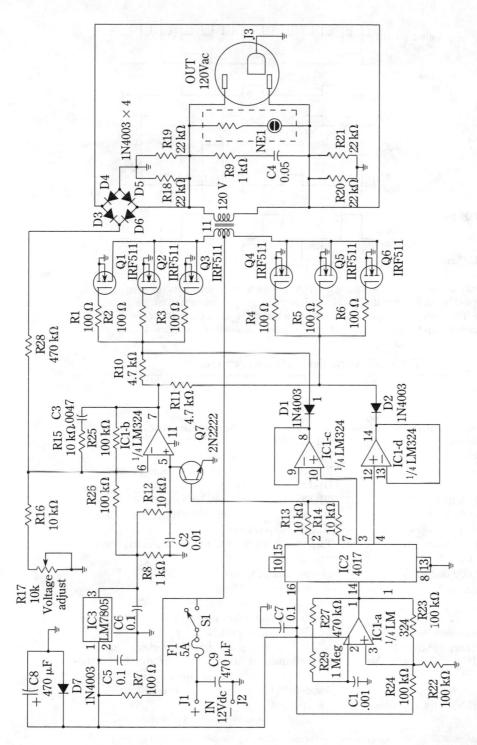

6-30 Schematic diagram of step-wave inverter. High efficiency is attained, and the output waveform is better for operating motors than a square wave. Greater power can be realized by paralleling more power-MOSFETS, or by using larger ones. Radio Electronics, April 1991, Gernsback Publications, Inc.

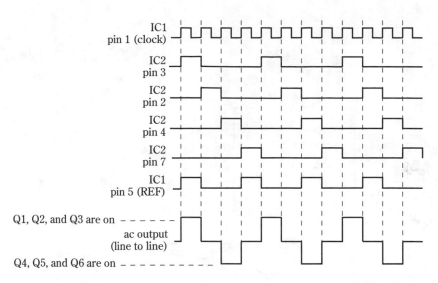

6-31 Waveform diagram of step-wave inverter. Pulses from a decade counter are sampled in such a way that divide-by-four triggering of the output-stage occurs. The resultant stepped wave is made up of 90° segments, and closely simulates a sine wave for motor operation. Radio Electronics, April 1991, Gernsback Publications, Inc.

Table 6-3. Parts list of step-wave inverter.

PARTS LIST

All resistors are ¼-watt, 5% unless otherwise noted.
R1–R7—100 ohms
R8–1000 ohms
R9–1000 ohms, ½-watt
R10, R11—4700 ohms
R12–R16—10,000 ohms
R17—10,000-ohm potentiometer
R18–R21—22,000 ohms, ½-watt
R22–R26—100,000 ohms
R27–R28—470,000 ohms
R29—1 megohm

Capacitors
C1—0.001 µF, ceramic disc
C2—0.01 µF, ceramic disc
C3—0.0047 µF, ceramic disc
C4—0.05 µF, 200 volts, ceramic disc or metal film
C5–C7—0.1 µF, ceramic disc
C8–C9—470 µF, 35 volts, electrolytic

Semiconductors
IC1—LM324 quad op-amp
IC2—4017 CMOS decade counter
IC3—LM7805 or LM340-5 +5-volt regulator
D1–D7—1N4003 diode
Q1–Q6—IRF511 60-volt 3.5-amp MOSFET
Q7—2N2222 or 2N3904 NPN transistor

Other components
T1—120/12.6 volt center-tapped 3-amp power transformer
J1—banana jack, red
J2—banana jack, black
J3—ac power receptacle
F1—5-amp slow-blow fuse
S1—SPST 6-amp switch
NE1—neon indicator light with series resistor

Miscellaneous: fuse holder, perforated construction board, enclosure aluminum for heatsinks, standoffs for mounting circuit board, wire, solder, etc

Radio-Electronics, April 1991, Gernsback Publications, Inc.

SubstituteIFRZ30 power-MOSFETS for higher output power (up to 250 watts). Use adequate heatsinks. Also substitute a Jefferson #216-1121 "buck-boost" transformer for T1. Connect 120-volt windings in *parallel*; connect 12-volt windings in *series*. Pay attention to phasing of these dual windings.

7
The nonclassic motors

ON THE ONE HAND, THE ENSUING DISCUSSIONS WILL APPEAR AS SMOOTH AND LOGICAL extensions of previously covered material. On the other hand, those who have become accustomed to associate certain performance parameters of motors with specific machine designs might not find it "natural" to adopt a new mind set in which the *control IC* merits most attention. Indeed, in the latest phase of the evolving technology, it appears that the control IC is the high-priority device. Once the motor fulfills the basic requirement of converting electrical energy to mechanical energy, its role becomes more suggestive of an auxiliary components.

Interestingly, many of the newer techniques substitute digital-drive impulses in place of dc or sinusoidal ac excitation. This leads to dramatic improvements in timing, precision, and repeatability. Commutation, rectification, and the production of rotating fields are accomplished in silicon "brains."

This being the case, the traditional electric-motor texts are likely to prove inadequate in dealing with newer applications. A surprising aspect of these advanced motor systems is that serious consideration is under way for integral horsepower designs. In industry, these would vie for applications so long monopolized by three-phase induction motors. And, in electric vehicles, such motor systems would be particularly easy to tailor performance characteristics for optimum speed, efficiency, and range requirements.

The off-the-beaten-path motor concept

The classic motors described in this book were designed specifically to be either dc or ac machines. Various operating characteristics ensued depending on whether a commutator and brushes were used, on the presence and connection of field windings, on the spatial distribution of windings, on the number of poles, and so on. Even after one complied with the basic laws of electromagnetism, much empirical work was necessary to accomplish trade offs that yielded a "civilized" motor—one with adequate starting torque, but not with excessive starting current; one that could run smoothly at slow speeds; one that was quickly responsive to both frictional and inertial load changes; one that could be conveniently braked, etc.

When electronic controls were first associated with motors, it was often unnecessary to select a particular type of motor known to excel in a desired performance characteristic. To a considerable extent, any of the speed/torque characteristics could be obtained from a variety of motors if the feedback logic of the control circuit was appropriate. For example, a near constant speed vs. variable load characteristic could be obtained with dc series or shunt motors, or with ac universal or induction motors. A synchronous motor could be operated as a variable-speed machine despite its reputation as a constant-speed machine. And stepping motors could be scaled up in size and used to drive heavy loads with a variety of programmed characteristics. Then, stepping motors equipped with Hall-effect or optical position sensors gave rise to designs known as brushless dc motors. This gave further impetus for semiconductor firms to develop more sophisticated control ICs. These were endowed with a variety of programmable options so that the user could tailor motor performance to his or her needs.

This situation is illustrated by the fantasized motor structure of Fig. 7-1. With minor variations, this prototype motor can be operated as a conventional synchronous motor, as a stepping motor, or as a brushless dc motor. Such variations include the use of rotor position sensors, and perhaps a damper winding on the rotor to promote self-starting and to suppress hunting. However, a commutator and brushes are *not* needed; commutation, timing, and sequencing logic are the job of the control IC. Indeed, the line of demarcation between "alternating current" and "direct current" motors becomes, at best, fuzzy. Among its myriad of functions, the control IC can perform the roll of inverter. In so doing, it also frees performance from the fixed-speed constraint imposed by conventional ac sources. All in all, motor technology is becoming based on silicon as well as iron and copper. And, although thyristors, Darlingtons, MOSFETs and bipolar transistors continue to provide the *muscle* of motor systems, the dedicated control IC will increasingly assume the roll of a brain.

The permanent-magnet stepping motor with unipolar drive

The basic principle of the stepper motor is that a rotor with either permanent or momentary magnetic poles is caused to turn in response to an incrementally rotating field pattern provided by stator windings. The action is suggestive of that in an "analog" synchronous motor in which a magnetized rotor also turns in response to a rotating field supplied by stator windings. However, the rotation of the field in the stepping motor occurs in digital jumps instead of smoothly or sinusoidally, as in the synchronous motor. Interestingly, the all-around performance of the stepping motor generally improves as these jumps become more numerous and closer together. The stepping motor operates from dc pulses that are timed and conditioned by a dedicated control IC (synchronous motors are increasingly being controlled by dedicated ICs, but they are fed polyphase ac drive).

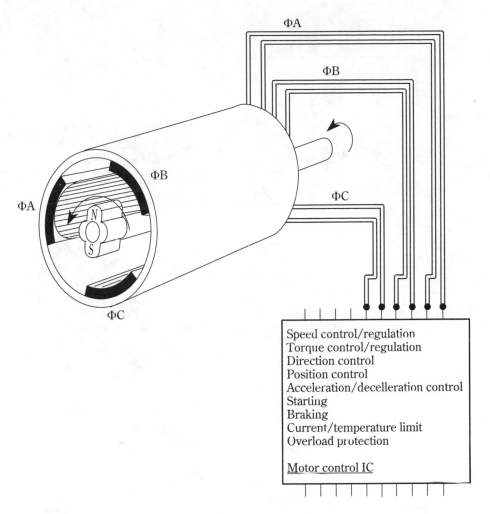

7-1 The "brains" of new-age motor design is in a control IC. Commutation, timing, and speed/torque characteristics are determined by the logic of a dedicated control-IC.

There are various ways of arranging and driving the stator windings of stepper motors, but two basic formats dominate. In the *unipolar* drive technique, such as that shown in Fig. 7-2, center-tapped or bifilar windings are used. Each pole pair accounts for six leads being brought out. One *or* the other of the windings comprising a bifilar pair is energized at any time. In this way, the magnetic poles generated by the stator windings can be selectively changed. The ground symbols in the sketch of Fig. 7-2 indicate the interconnections of bifilar winding pairs that constitute the center taps. In practice these center taps also connect to the power-supply ground, either directly or through a current-limiting resistance. In any event, phase voltages applied to the remaining four leads are either on or off.

Additional insight into the operation of the unipolar-type stepping motor can be gleaned from the functional diagram of Fig. 7-3. The switching logic needed to obtain

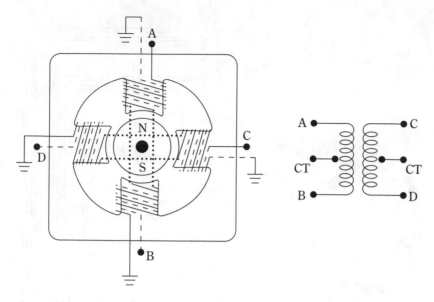

7-2 Permanent-magnet stepping motor with unipolar windings. The drive circuit operates from a fixed-polarity dc source. Stator poles are either N or S in response to *which* of the bifilar windings is energized. Single-ended drivers are generally used to turn winding currents on and off.

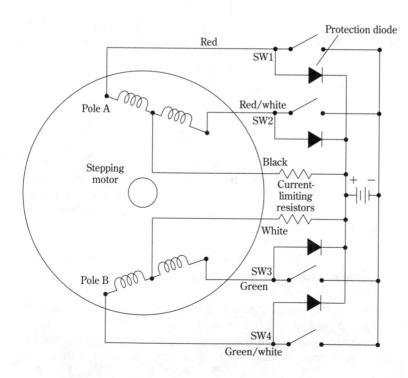

7-3 Functional diagram of unipolar-winding stepping motor and drive. The drive comprises four electronic-switches and a dc power-source. Current direction is never reversed. Phase sequence advances via *selection* of appropriate windings. Full-step and half-step rotational increments are readily attained from the switching logic.
Motorola Semiconductor Products, Inc.

either full-step or half-step operation is indicated in Table 7-1A. The half-step mode tends to give smoother, less noisy operation with higher torque and efficiency. The reasons are not unlike those pertaining to a comparison between an eight-cylinder and a four-cylinder automotive engine. In both cases, there are more "power strokes" per revolution. Special ICs are available for dividing the stepping increments even further so that there are a great many stops per revolution. This is known as microstepping control and yields the ultimate smoothness of operation. In this way, the continuous torque characteristic of linear motors can be approached.

Table 7-1A.
Switching logic for the simple
unipolar stepping motor. Reverse
rotation is brought about by reversing
the phase sequence of the switching states.

		Full-step sequence		
Step	**SW1**	**SW2**	**SW3**	**SW4**
1	OFF	ON	OFF	ON
2	OFF	ON	ON	OFF
3	ON	OFF	ON	OFF
4	ON	OFF	OFF	ON
1	OFF	ON	OFF	ON

		Half-step sequence		
Step	**SW1**	**SW2**	**SW3**	**SW4**
1	OFF	ON	OFF	ON
2	OFF	ON	OFF	OFF
3	OFF	ON	ON	OFF
4	OFF	OFF	ON	OFF
5	ON	OFF	ON	OFF
6	ON	OFF	OFF	OFF
7	ON	OFF	OFF	ON
8	OFF	OFF	OFF	ON
1	OFF	ON	OFF	ON

An example of a stepping-motor system with unipolar drive is shown in Fig. 7-4. As previously explained, the stator windings are arranged with innerjunction or center-tap leads brought out; these are the black and white leads so designated in Fig. 7-4. Instead of being grounded in this application, these two leads connect to +24 V through current-limiting resistances *R1* and *R2*. The other winding leads are connected to individual MOSFET switching stages that determine whether the current path of each winding is completed to ground.

In the circuit of Fig. 7-4, the deployment of the two MC14194 four-bit shift registers provides the eight unique switching conditions needed for the most demanding stepping logic, which is the half-step sequence. (Full-step sequence can be programmed by setting the half-step line at logic high and performing a preset).

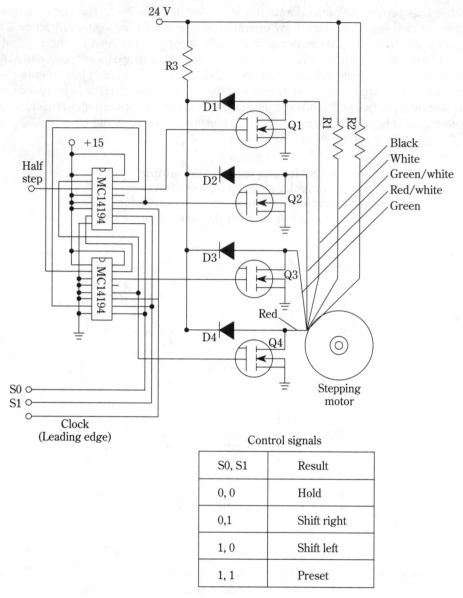

S0, S1	Result
0, 0	Hold
0, 1	Shift right
1, 0	Shift left
1, 1	Preset

Control signals

Parts:
1. Example motor is Superior Electric's
   Model MO93-FC07
2. Diodes, D1–D4, 1N4002 or equivalent
3. R3, 10 Ω 10 W
4. R1, R2, 6.5 Ω 100 W
5. Integrated circuits, MC14194 (2 required)
6. Q1–Q4, MTM12N08 or MTP12N08

7-4 Example of stepping motor drive with center-tapped windings. This is a *unipolar* system. Note standard color-code of winding leads. Either full-step or half-step operation can be obtained. Motorola Semiconductor Products, Inc.

Right shifting causes one direction of rotation; left shifting reverses the direction at rotation. Diodes 1–4 protect the power MOSFETs from inductive turn-off spikes. Capacitors on the order of 0.03 μF (not shown) connected across the windings can provide additional protection from the switching transients.

Although resistances R_1 and R_2 are basically used for current limiting, they also serve *another* function. Because these resistances lower the L/R time constant of the motor winding circuits, they enable higher stepping rates to be attained than would be the case otherwise. Indeed, even higher stepping rates will result from increasing the values of these resistances and simultaneously increasing the dc supply voltage. A disadvantage of this technique is that the power dissipation in these resistances lowers the overall efficiency of the drive system.

Another method for increasing torque at high stepping rates can be implemented that does not waste power as does the above method. Remember, the basic idea is to hasten the rise of current in the motor windings, inasmuch as torque is proportional to drive current. As inferred, when the stepping rate is high, there is insufficient time for optimum current rise to take place. This is true despite the motor designer's goal to keep winding inductance as low as possible. Raising the dc supply voltage causes a faster rise of current in the motor windings, but it is necessary that the average current does not increase beyond safe levels. The insertion of resistance, as pointed out, wastes power. An alternate way of keeping average current at safe levels is to switch this current at a high rate. In so doing, the objective of achieving a high rate of rise will be met. Also, because of the switching technique, the average current level in the windings can be prevented from becoming unsafe. All the while, torque will be increased without incurring needless power dissipation. These benefits will be most realizable at high stepping rates (an example of current switching is shown following the discussion of bipolar-drive stepping motors).

The permanent-magnet stepping motor with bipolar drive

A permanent-magnet stepping motor with *bipolar* windings is illustrated in Fig. 7-5. With this winding configuration on the stator, the driver must be able to send current in either direction through a selected pair of windings so that the resultant magnetic field experienced by the rotor can be either N-S or S-N in direction. This calls for push-pull or H-bridge drivers. Other things being equal, stepping motors with bipolar windings tend to exhibit more precise operating characteristics than do stepping motors with unipolar windings. This is because the polarity reversals experienced by the bipolar windings are effective in quickly neutralizing the stored energy in the windings. By contrast, the unipolar type of motor tends to have a "mushy" running characteristic. However, it is also true on an "other things being equal basis," that the unipolar winding format allows higher torque at higher stepping rates. It is fitting to state that an earlier objection to the bipolar type of motor has been largely overcome. This pertains to the requirement for more complex drive electronics. This need is now easily met by control ICs.

Three commonly used driving formats are used for bipolar stepping motors. These are shown in the diagram of Figs. 7-6, 7-7, and 7-8. In the "one-phase-on"

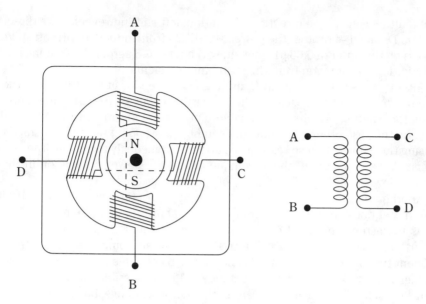

7-5 Permanent-magnet stepping motor with bipolar windings. The drive circuit can send current through the windings in *either* direction so that the stator poles can be either N or S. Complementary symmetry, push-pull, or H-bridge drivers can be used.

method of Fig. 7-6, full-step rotational increments are obtained. Because only one phase of the stator windings is energized at a time, a penalty in torque must be acceptable when this drive sequence is used.

In the two-phase-on drive method shown in Fig. 7-7, full-step increments of rotor rotation are also obtained, but the rotor now lines up at the inter-phase positions. Higher torque is available than from the previous method because now two phases contribute to the stepping process.

As might be expected, the two described phase sequence can be combined to produce the half-step drive technique shown in Fig. 7-8. This is brought about by alternating between the full-step drive methods as the stepping process sequences. Naturally, there are now eight, rather than four, unique states per revolution.

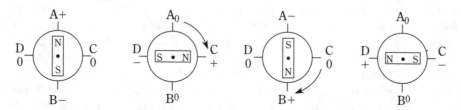

7-6 The one-phase-on method of driving a permanent-magnet stepping motor. Note that only *one* phase at a time is energized. The rotor turns in full-stop increments in order to line up with the magnetic field of the stator windings.

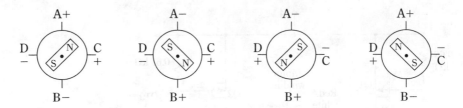

7-7 The two-phase-on method of driving a permanent-magnet stepping motor. This phase sequence produces higher torque than the one-phase-on method. The rotor, however, still turns in one-step increments. Note that the two phases are always *both* energized to produce rotation.

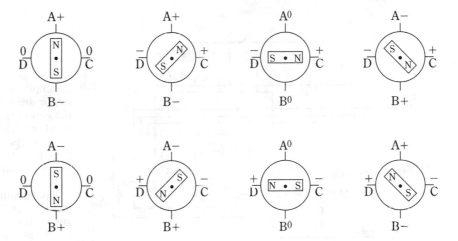

7-8 The half-step method of driving a permanent-magnet stepping motor. Here, the one-phase-on and the two-phase-on driving methods are combined. The phase sequence, or "firing order", alternates between the two methods. The net result is that the rotor turns in half-step increments.

In order to gain finer resolution and smoother torque, real-life stepping motors might have multiple rotor poles, rather than the two depicted in the diagram. An even number of permanent-magnet poles are always used. This statement might appear self-evident, but there is another type of stepping motor in which the actual number of rotor poles is not what might be expected from a "common sense" approach.

An example of a stepping-motor system utilizing bipolar windings is shown in Fig. 7-9. The same stepping motor can be used as in the unipolar circuit of Fig. 7-4, but the winding center taps are then left unconnected. As in Fig. 7-4, two shift-register ICs provided the stepping logic for either half-, or full-step operation. In this circuit, current limiting for the stator windings is provided by the two 9-Ω resistances. *Two* of each component and device depicted in the driver portions of the circuit of Fig. 7-9 are needed because this circuit is drawn for only one stator winding. How-

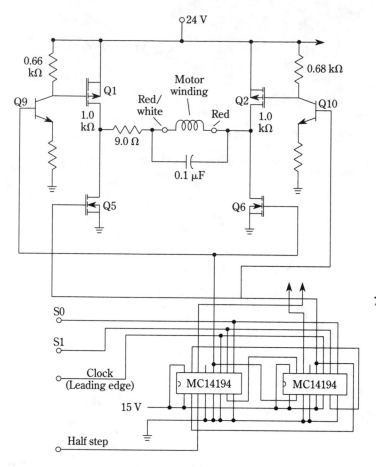

7-9 Example of stepping motor drive with noncenter-tapped windings. This is a bipolar system. The complete logic for a two-phase motor is shown. Only the push-pull complementary-symmetry drive for *one* phase is shown for simplicity; identical drive circuitry is used for the two phases.

Motorola Semiconductor Products, Inc.

S0, S1	Result
0, 0	Hold
0, 1	Shift right
1, 0	Shift left
1, 1	Preset

Parts:
1. 2 × MC14194 (CMOS)
2. Q1–Q4, 4 × MTP8P08 or MTM8P08 (TMOS power FETs)
3. Q5–Q8, 4 × MTP12N08 or MTM12N08 (TMOS power FETs)
4. Q9–Q12, 4 × MPS8099 (NPN small signal)
5. 2 × 9 Ω 50-W resistors
6. 4 × 0.68 kΩ ¼-W resistors
7. 4 × 1.0 kΩ ¼-W resistors
Example motor: Superior Electric SLO-SYN model M093-F007

ever, the logic suffices, as shown, for both motor windings. Specifically, the requirement in Fig. 7-9 is for four N-channel power MOSFETs, four P-channel power MOSFETs, but only two MC14194 ICs.

The parameters of a full winding of the stepping motor used in the previous examples are as follows: the resistance is 2.5 Ω; inductance is 31.8 μH, and rated current is 2.0 A. The designer of these motors must trade off the conflicting needs imposed by torque, magnetic saturation, temperature rise, and winding space; he or she then hopes the inductance will be as low as possible so that the torque will hold up for reasonably high stepping rates. In practice, circuit and operating techniques might have to be resorted to overcome the current-retarding effect of winding inductance. The bottom line is that a high stepping rate requires a high d_i/d_t in the motor windings. The user of the motor can increase the d_i/d_t by raising the dc supply voltage and then relying either on series resistance, or on a switching technique, to keep the average winding currents at rated levels.

What appears to be a resonating capacitor across the motor winding is not large enough to function in that manner within frequency bands involved in the operation. Indeed, both electrical and mechanical resonances tend to be very undesirable. This capacitor absorbs inductive-kickback energy, thereby protecting the power MOSFETs from avalanching. This protective function is supposed to be the entire job of the internal body diodes of the MOSFETs, but in practice they do not turn on instantaneously; it requires about 300 nanoseconds for these diodes to become fully conductive.

Some semiconductor firms market power ICs comprised of four power MOSFETs designed to function as drivers of bipolar stepping motors. This keeps the parts count down and economizes on production costs.

Drive-current switching for improving speed, torque, and efficiency of stepping motors

Inspect the circuitry of Fig. 7-10. Notice that the logic, as well as the power MOSFET switching bridge, remain essentially the same as in the circuit of Fig. 7-9. There have obviously been some circuit *additions* made in Fig. 7-10, but what is significant is what has been left out. Specifically, the 50-W, 9-Ω current-limiting (and power-wasting) resistance has been omitted in Fig. 7-10. It is no longer needed because current limiting is now accomplished by high-frequency interruption of the drive circuit. This is introduced by the two comparators that turn the MOSFET bridge off and on in response to preset current levels sensed as voltage drops across the 0.1-Ω resistances. As previously mentioned, this oscillatory mode enables fast current rise during the phase sequence of the motor, but maintains the average current in the windings at a safe level. To advantageously use this switching technique, the dc supply voltage is increased to 70 V.

The schematics of both Fig. 7-9 and Fig. 7-10 only show the driver circuitry for *one* motor winding, it being understood that duplicate drivers are needed for the other motor winding. The "brains" or logic, however, suffice as shown, for both motor windings. By inspecting the parts list of Table 7-1B, you can quickly determine which circuit components must be had in duplicate, and which individually serve the needs of the entire motor.

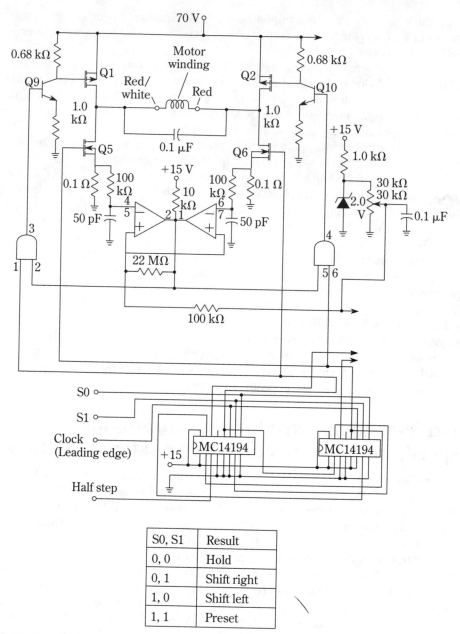

7-10 Switched-current drive applied to the stepping motor system of Fig. 7-9. Oscillatory interruption of the winding-current limits it to a safe value, but allows it to rise rapidly. This enhances torque at high stepping-rates. Unlike the resistive method of Fig. 7-9, power dissipation is very low. Motorola Semiconductor Products, Inc.

**Table 7-1B. List of parts for the
switched-current drive-system of Fig. 7-10.**

Integrated circuits
1. 2 × MC14194B, CMOS 4-bit shift register
2. 1 × MC14081B, CMOS Quad "AND" gate
3. 1 × MLM399P, Quad comparator

TMOS power FETs
1. Q1–Q4, 4 × MTP8P08 or MTM8P08, P-channel power FET
2. Q5–Q8, 4 × MTP12N08 or MTM12N08, N-channel power FET

Transistors
1. Q9–Q12, 4 × MPS8099, NPN small signal transistors

Resistors
1. 4 × 0.1 Ω 2.0 W
2. 4 × 680 Ω ¼ W
3. 5 × 1.0 kΩ ¼ W
4. 2 × 10 kΩ ¼ W
5. 1 × 30 kΩ ¼ W
6. 1 × 30 kΩ adjustable, ¼ W
7. 6 × 100 kΩ ¼ W
8. 2 × 22 MΩ ⅛ W

Zener diode
1. 1 × MZ4679, 2 V reference

Capacitors
1. 3 × 0.1 μF 100 V
2. 4 × 50 pF 50 V

Motorola Semiconductor Products, Inc.

The current-interruption rate generated by the comparator circuitry is not of prime importance; it is only desirable that this rate be much higher than the stepping rate of the motor. The approximate current-rise speed will be V/L, where V is the supply voltage and L is the inductance of the driven motor winding. Although this does not take into account the ohmic resistance of the windings, it can be appreciated that drive current should rise about three times faster than when working from a 24-V supply. The practical consequence of these matters is that only 25 percent of supplied power gets to the motor in the setup of Fig. 7-9, whereas the switching technique used in Fig. 7-10 allows nearly 80 percent of supplied power to drive the motor. The comparison is fair inasmuch as the cited efficiencies apply for per-winding currents of two amperes in both cases.

As a practical hint, it should prove expedient to get the simpler circuit of Fig. 7-9 operative first. Then, the sophistications depicted in Fig. 7-10 can be added. In this way, troubleshooting, if required, is likely to be more straightforward.

The variable-reluctance stepping motor

The second type of stepping motor is the variable-reluctance structure in which the rotor is made of nonmagnetized soft iron. Such a machine is illustrated in Fig. 7-11. The salient operating feature is that permanent rotor poles do *not* develop. Physically, there are other differences from the permanent-magnet stepping motor. What might initially appear to be error on the part of a not-too-alert draftsman is actually representative of this type of variable-reluctance stepping motor—there are *fewer* rotor teeth than stator poles. Also, the stator poles have nonuniform widths. These design features help produce steady torque throughout the rotation. These stepping motors are generally equipped with unipolar stator windings, but center taps are not necessary.

At first consideration, it might be a brainteaser to determine the number of degrees the rotor must turn for a full step of rotation. Actually, a similar calculation is used as with permanent-magnet stepping motors. Simply divide 360° by the product of the rotor teeth and the number of pairs of stator windings (each pair of stator windings constitutes a *phase*). Reduced to numbers, you have

$$\frac{360}{8} \times 4$$

or 15°. Do not be confused by the momentary life span of the magnetic poles induced in the soft-iron rotor, or by the unequal number of poles in the stator and the rotor. Note also that 15° is the difference in angular pitch between stator and rotor teeth. That is, 60° − 45° = 15°.

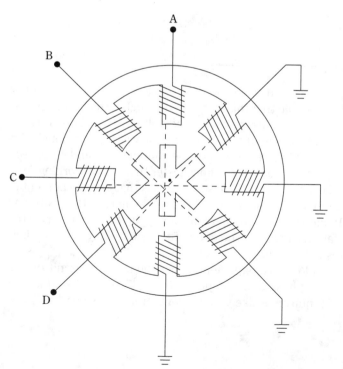

7-11 A variable-reluctance stepping motor. Torque production requires certain departures from the structures used in the permanent-magnet stepping motor. Note the variance in the stator poles. Also, it is seen that there are *fewer* rotor than stator poles. Unipolar windings are used in this example.

The drive logic of the variable-reluctance stepping motor is similar to that of the permanent-magnet types. Because unipolar windings are used on the stator, there are no polarity-reversals in the drive pulses. The three optional phase sequences for producing rotation are as follows: The one-phase on, or wave sequence, is simply *A/B/C/D*. The two-phase on sequence is *AC/CB/BD/DA*. These are both full-step sequences. They differ by a half step in the angular displacement of the rotor. By combining these two full-step sequences, you obtain the half-step sequence, *A/AC/C/BC/B/BD/D/DA*.

An important difference in behavior between permanent-magnet stepping motors and the reluctance type is that the latter do not posses *detente torque*, sometimes called "position memory." That is, the permanent-magnet type, when unenergized, will nonetheless exhibit resistance to movement. In contrast, the unenergized reluctance stepping motor "forgets" its most recently energized position.

There are more complicated variable-reluctance designs than that of Fig. 7-11. A common structural scheme comprises three ganged rotors all locked to the common shaft. Each such rotor has the same number of teeth as a stator concentric with it. However, each of the rotors is slightly staggered so that the rotor teeth do not line up. The stator structures have their individual windings. By supplying appropriately sequenced excitation to the three stator windings, *one* of the rotors will turn in order to attain a position of minimum reluctance. In this way, the motor develops incremental steps of torque, and continuous rotation can be produced.

Electronic commutation for the dc brushless motor

If ever a single type of motor will be found to have characteristics suitable for nearly every application serviced by the many ac and dc motors traditionally used, it will likely be the *dc brushless motor*. In a way, this machine can be viewed as a dc motor in which electronic commutation replaces the commutator and brushes. The mere fact that the commutator-brush assemblage is dispensed with, implies low maintenance, low RFI and EMI, and low cost. But the electronic commutation provided by a dedicated IC does even more; it enables convenient "tailoring" of motor characteristics so that compromises that formerly attended the selection of a motor type need not apply. For example, such features as high starting torque and variable speed can now easily be realized together. And previous debates about the use of a dc or an ac motor tend to lose their relevance when a brushless dc motor is used with its accompanying dedicated IC.

One such dedicated IC is shown in the block diagram of Fig. 7-12. This IC is essentially concerned with electronic commutation; speed, torque, and direction control would be accomplished via the association of a separate PWM IC. Although the LM621 brushless motor commutator does not have on-board PWM logic, there are other dedicated ICs that do. Advantages can be cited for both architectures. Having the commutation and control functions on separate ICs probably allows more flexibility in achieving specialized performance features. The pinout functions of the IM621 are listed in Table 7-2.

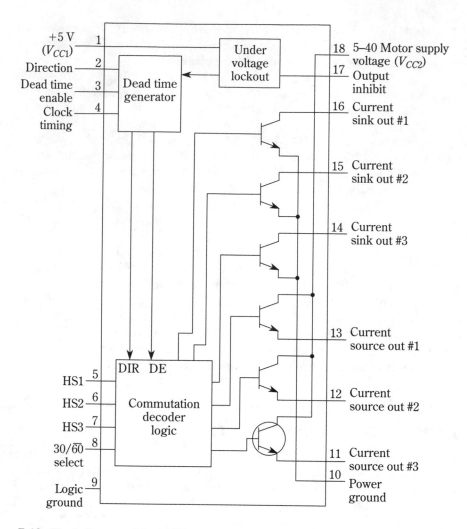

7-12 Block diagram of the LM621 brushless dc motor commutator. This dedicated-IC enables mechanically-simple dc motors with no physical commutators or brushes to be driven from a dc source. National Semiconductor Corp.

Table 7-2. The pinout functions of the LM621 brushless dc motor commutator. The two common types of three-phase motors (30°and 60°), as well as four-phase motors (90°) can be accommodated.

Pin 1: V_{CC1} (+5 V). The logic and clock power supply pin.

Pin 2: DIRECTION. This input determines the direction of rotation of the motor; i.e., clockwise vs. counterclockwise. See truth table.

Pin 3: DEAD-TIME ENABLE. This input enables or disables the dead-time feature. Connecting +5 V to pin 3 enables dead-time, and grounding pin 3 disables it. Pin 3 should not be allowed to float.

Pin 4: CLOCK TIMING. An RC network connected between this pin and ground sets the period of the clock oscillator, which determines the amount of dead-time. See Fig. 2 and text.

Pin 5 thru 7: HS1, HS2, and HS3 (Hall-sensor inputs). These inputs receive the rotor-position sensor inputs from the motor. Three-phase motors provide all three signals; four-phase motors provide only two, one of which is connected to both HS2 and HS3.

Pin 8: 30/60 SELECT. This input is used to select the required decoding for three-phase motors; i.e., either "30-degree" (+5 V) or "60-degree" (ground). Connect pin 8 to +5 V when using a four-phase motor.

Pin 9: LOGIC GROUND. Ground for the logic power supply.

Pin 10: POWER GROUND. Ground for the output buffer supply.

Pins 11 thru 13: SOURCE OUTPUTS. The three current-sourcing outputs which drive the external power devices that drive the motor.

Pins 14 thru 16: SINK OUTPUTS. The three current-sinking outputs which drive the external power devices that drive the motor.

Pin 17: OUTPUT INHIBIT. This input disables the LM621 outputs. It is typically driven by the magnitude signal from an external sign/magnitude PWM generator. Pin 17 = +5 V = outputs off.

Pin 18: V_{CC2} (+5 to +40 V). This is the supply for the collectors of the three current-sourcing outputs (pins 11 thru 13). When driving MOSFET power devices, pin 18 may be connected to a voltage source of up to +40 V to achieve sufficient output swing for the gate. When driving bipolar power devices, pin 18 should be connected to +5 V to minimize on-chip power dissipation. Undervoltage lockout automatically shuts down all outputs if the V_{CC1} supply is too low. All outputs will be off if V_{CC1} falls below the undervoltage lockout voltage.

National Semiconductor Corp.

As is commonplace with ICs intended for motor-drive purposes, important protective features "come along for the ride". One of these is the adjustable dead-time feature, which eliminates current spikes that can be especially destructive when the motor is being braked or reversed. An undervoltage-lockout provision inhibits motor drive in the event of low motor supply voltage. The outputs can accommodate the gate-voltage swings for driving power MOSFETs, or can deliver 35 milliamps to each base of six bipolar power transistors.

Additionally, the LM621 is compatible with the drive requirements of both three- and four-phase dc brushless motors. Both, the 30° sensor displacement and the 60° sensor displacement types of three-phase motors can be handled. These accommodations are depicted in the commutator decoder truth table Table 7-3.

**Table 7-3. Commutator decoder truth table for the
LM621 brushless-motor commutator. Table pertains to the
two common types of three-phase motors, as well as to four-phase motors.**

Sensor phasing	Position range	Sensor inputs			Sink outputs			Source outputs		
		HS1	HS2	HS3	1	2	3	1	2	3
	0–60	0	0	0	ON	off	off	off	ON	off
	60–120	0	0	1	ON	off	off	off	off	ON
30 deg	120–180	0	1	1	off	ON	off	off	off	ON
	180–240	1	1	1	off	ON	off	ON	off	off
	240–300	1	1	0	off	off	ON	ON	off	off
	300–360	1	0	0	off	off	ON	off	ON	off
	0–60	1	0	1	ON	off	off	off	ON	off

Table 7-3. Continued

Sensor Position		Sensor inputs			Sink outputs			Source outputs		
phasing	range	HS1	HS2	HS3	1	2	3	1	2	3
	60–120	1	0	0	ON	off	off	off	off	ON
60 deg	120–180	1	1	0	off	ON	off	off	off	ON
	180–240	0	1	0	off	ON	off	ON	off	off
	240–300	0	1	1	off	off	ON	ON	off	off
	300–360	0	0	1	off	off	ON	off	ON	off
	0–90	0	1	HS2	off	na	off	off	na	ON
90 deg	90–180	0	0	HS2	ON	na	off	off	na	off
	180–270	1	0	HS2	off	na	ON	off	na	off
	270–360	1	1	HS2	off	na	off	ON	na	off
Pin numbers:		5	6	7	16	15	14	13	12	11

Note 1: The above outputs are generated when the direction input, pin 2, is logic high. For reverse rotation (pin 2 logic low), the above sink and source output states become exchanged.

Note 2: For four-phase motors sink and source outputs number two (pins 15 and 12) are not used; hence the "na" (not applicable) in the appropriate columns above.

National Semiconductor Corp.

The basic circuit arrangement for electronically commutating a three-phase dc brushless motor with the LM621 IC is shown in Fig. 7-13. Six N-channel power MOS-FETs drive the motor's delta-connected stator windings. The Hall-effect rotor position sensors connect directly to pins 5, 6, and 7 of the LM621. The effective magnitude of motor drive current is controlled via the duty cycle of pulses from an *external* pulse-width modulator circuit. This is tantamount to saying that motor speed is controlled by the pulse-width modulator. This is a nice technique of speed control because motor torque is not sacrificed at low speeds; the practical result is that a wide range of speeds can be obtained. Also, the low-speed cogging common to conventional dc motors is absent.

The PWM frequency is not critical; it will probably be in the audio or low super-sonic region and can be determined empirically for best results. The size and winding characteristics of the motor will tend to govern this parameter. Originally, this motor control technique was applied to small motors, but successful applications have been made with fractional-horsepower motors. It now appears that dc brushless motors of integral-horsepower capability can be efficiently controlled with electronic commutation and pulse-width modulation. Perhaps this technique can be extended for application to electric vehicles. If so, the ongoing controversy between dc and ac motors will end, because we will then have a motor combining the most desirable characteristics of *both* types.

The commutation waveform diagram for the motor-driven system of Fig. 7-13 is shown in Fig. 7-14. Notice that reversal of the direction of rotation is brought about by reversal of the sequence of drive currents to the stator windings. However a physical change of connections is not necessary; rather, reversal is accomplished via the logic sign applied to pin 2, which causes the source and sink output states to become exchanged.

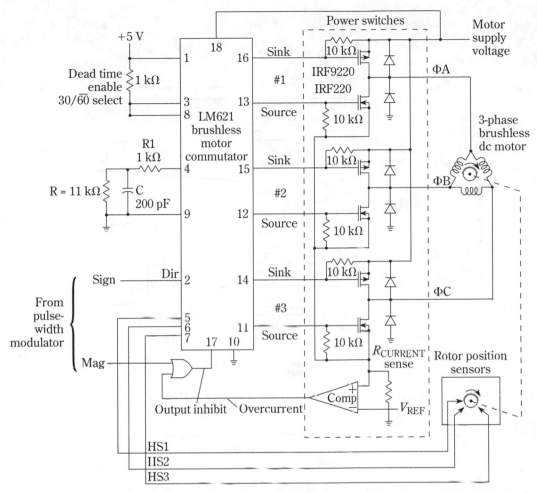

7-13 Commutator scheme for a three-phase dc motor. This is representative of a newly developing technology in which features of both dc and ac motors are combined in a single machine. Operation is from a dc source with no need of a mechanical brush-commutator system. National Semiconductor Corp.

You might have noticed that the architecture of this motor resembles that of a three-phase synchronous motor, which also has a permanent-magnet rotor and either Y or delta-connected stator windings. In the synchronous motor, speed is determined by the drive *frequency*. In this application of the dc brushless motor, the frequency of the PWM pulses have no bearing on motor speed, providing the duty cycle of these pulses is not changed. In other words, it is the effective drive *current* in the windings that governs speed. This mode of operation is due to the electronic commutation—in the ac synchronous motor, no such process interferes with the rotor's "desire" to follow the rotating field. Incidentally, a good frequency for the PWM pulses is about 25 kHz because it is above the audio range, but low enough to be faithfully reproduced by the switching transistors.

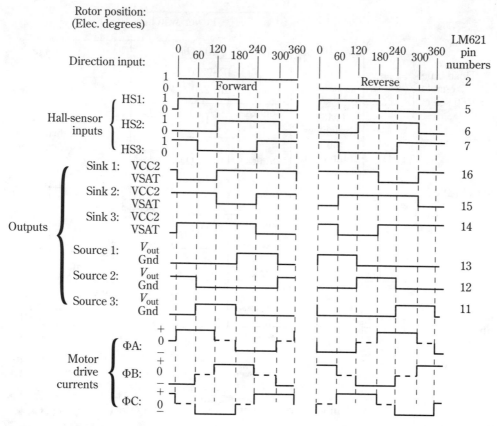

7-14 Commutator waveforms for dc brushless motor with three-phase stator. These waveforms pertain to machines in which the physical displacement between the Hall-effect sensors is 30°. National Semiconductor Corp.

Another manufacturer's version of electronic commutation for brushless dc motors conveniently has the pulse-width modulator circuitry *included* in the chip. The block diagram of this dedicated IC, the Motorola MC33035 is shown in Fig. 7-15. Notice that the mere addition of an external potentiometer enables speed control of the motor. Also, an external resistance and capacitor allow for selection of the oscillator frequency. A practical frequency would be in the 20 kHz to 25 kHz region. The pin connections are explained in Table 7-4.

Table 7-4. Description of pin-functions of the MC33035 brushless dc motor controller. This IC is available in 24-pin dual-in-line plastic packages.

Pin no.	Function	Description
1, 2, 24	B_T, A_T, C_T	These three open collector Top Drive Outputs are designed to drive the external upper power switch transistors.
3	FWD/REV	The Forward/Reverse Input is used to change the direction of motor rotation.

Pin no.	Function	Description
4, 5, 6	S_A, S_B, S_C	These three Sensor Inputs control the commutation sequence.
7	Output Enable	A logic high at this input causes the motor to run, while a low causes it to coast.
8	Reference Output	This output provides charging current for the oscillator timing capacitor C_T and a reference for the error amplifier. It may also serve to furnish sensor power.
9	Current Sense (Noninverting Input)	A 100 mV signal, with respect to pin 15, at this input terminates output switch conduction during a given oscillator cycle. This pin normally connects to the top side of the current sense resistor.
10	Oscillator	The Oscillator frequency is programmed by the values selected for the timing components, R_T and C_T.
11	Error Amp (Noninverting Input)	This input is normally connected to the speed set potentiometer.
12	Error Amp (Nonverting Input)	This input is normally connected to the Error Amp Output in open-loop applications.
13	Error Amp Output/ PWM Input	This pin is available for compensation in closed-loop applications.
14	$\overline{\text{Fault}}$ Output	This open collector output is active low during one or more of the following conditions: Invalid Sensor Input code, Enable Input at logic 0, Current Sense Input greater than 100 mV (pin 9 with respect to pin 15), Undervoltage Lockout activation, and Thermal Shutdown.
15	Current Sense (Inverting Input)	Reference pin for internal 100 mV threshold. This pin is normally connected to the bottom side of the current sense resistor.
16	Ground	This pin supplies a ground for the control circuit and should be referenced back to the power source ground.
17	V_{CC}	This pin is the positive supply of the control IC. The controller is functional over a minimum V_{CC} range of 10 V to 30 V.
18	V_C	The high state (V_{OH}) of the Bottom Drive Outputs is set by the voltage applied to this pin. The controller is operational over a minimum V_C range of 10 V to 30 V.
19, 20, 21	C_B, B_B, A_B	These three totem pole Bottom Drive Outputs are designed for direct drive of the external bottom power switch transistors.
22	60°/$\overline{120°}$ Select	The electrical state of this pin configures the control circuit operation for either 60° (high state) or 120° (low state) sensor electrical phasing inputs.
23	Brake Input	A logic low state at this input allows the motor to run, while a high state does not allow motor operation and if operating causes rapid deceleration.

Motorola Semiconductor Products, Inc.

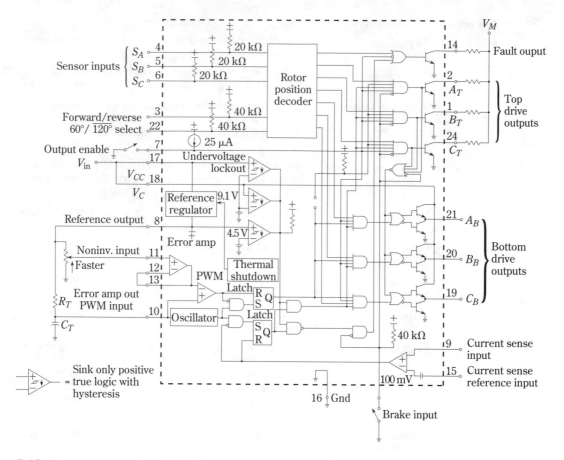

7-15 Block diagram of the MC33035 dc motor control LC. The self-contained PWM circuitry enables easy implementation in both, open-loop and in closed-loop applications. Motorola Semiconductor Products, Inc.

The circuit of a three-phase, six-step, full-wave motor control system is shown in Fig. 7-16 and the truth table for this system is shown in Table 7-5. Several practical differences from the previously discussed commutation arrangement are worthy of mention. This scheme shows a combination of bipolar power transistors and power MOSFETs as motor drivers (actually, the bipolar devices are Darlingtons). Fault-Output pin 14 enables an external LED to turn on in order to signify any of the fault conditions listed in Table 7-4. Activation of the LED is accompanied by deactivation of all motor-drive outputs. Note also, Brake Input pin 23, by which the motor can be rapidly decelerated. Such dynamic braking takes place when pin 23 is put in a high state; this causes the top-drive outputs to turn off and the bottom drives to turn on, shorting the motor-generated counter EMF. Brake input has unconditional priority over all other inputs. The bottom-drive power devices, power MOSFETs in this case, have to be selected with ample current capability.

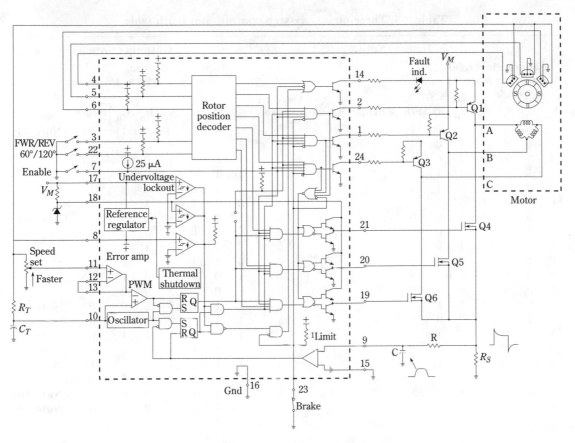

7-16   Three-phase, six-step, full-wave motor control system. The MC33035 IC provides electronic commutation for the three-phase brushless dc motor. Open-loop control of speed is accomplished by adjusting the duty cycle of the three bottom power-MOSFETS. Motorola Semiconductor Products, Inc.

All told, these differences from the previously discussed commutation system do not constitute drastic variances. Both company's control schemes are of the open-loop type. Closed-loop regulation of speed can be implemented in both with the same design philosophy as used in PWM voltage-regulated power supplies. That is, a sampled-voltage representing motor speed is introduced at the input of the error amplifier, the output of which causes duty-cycle variation from PWM circuitry; the power-boosted pulses are then applied to the motor and counteract any tendency for it to change its speed. This is particularly straightforward with the MC33035 IC, where the output of a motor-mounted dc tachometer can be applied to pin 12 of the error amplifier. An alternate method is to use the pulsed output of one of the Hall-effect sensors to provide the speed information to the error amplifier. The circuit implementation of this method is shown in Fig. 7-17. A similar implementation using a dedicated IC for greater precision is shown in Fig. 7-18.

Table 7-5. Three-phase, six-step commutator truth table.

Inputs (Note 2)										Outputs (Note 3)							
Sensor electrical phasing (Note 4)										Top drives			Bottom drives				
60°			120°						Current							$\overline{\text{Fault}}$	
S_A	S_B	S_C	S_A	S_B	S_C	F/R	Enable	Brake	sense	A_T	B_T	C_T	A_B	B_B	C_B	Fault	
1	0	0	1	0	0	1	1	0	0	0	1	1	0	0	1	1	(Note 5)
1	1	0	1	1	0	1	1	0	0	1	0	1	0	0	1	1	F/R = 1
1	1	1	0	1	0	1	1	0	0	1	0	1	1	0	0	1	
0	1	1	0	1	1	1	1	0	0	1	1	0	1	0	0	1	
0	0	1	0	0	1	1	1	0	0	1	1	0	0	1	0	1	
0	0	0	1	0	1	1	1	0	0	0	1	1	0	1	0	1	
1	0	0	1	0	0	0	1	0	0	1	1	0	1	0	0	1	(Note 5)
1	1	0	1	1	0	0	1	0	0	1	1	0	0	1	0	1	F/R = 0
1	1	1	0	1	0	0	1	0	0	0	1	1	0	1	0	1	
0	1	1	0	1	1	0	1	0	0	0	1	1	0	0	1	1	
0	0	1	0	0	1	0	1	0	0	1	0	1	0	0	1	1	
0	0	0	1	0	1	0	1	0	0	1	0	1	1	0	0	1	
1	0	1	1	1	1	X	X	0	X	1	1	1	0	0	0	0	(Note 6)
0	1	0	0	0	0	X	X	0	X	1	1	1	0	0	0	0	Brake = 0
1	0	1	1	1	1	X	X	1	X	1	1	1	1	1	1	0	(Note 7)
0	1	0	0	0	0	X	X	1	X	1	1	1	1	1	1	0	Brake = 0
V	V	V	V	V	V	X	1	1	X	1	1	1	1	1	1	1	(Note 8)
V	V	V	V	V	V	X	0	1	X	1	1	1	1	1	1	0	(Note 9)
V	V	V	V	V	V	X	0	0	X	1	1	1	0	0	0	0	(Note 10)
V	V	V	V	V	V	X	1	0	1	1	1	1	0	0	0	0	(Note 11)

NOTES:

1. V = Any one of six valid sensor or drive combinations
   X = Don't care.

2. The digital inputs (Pins 3, 4, 5, 6, 7, 22, 23) are all TTL compatible. The current sense input (Pin 9) has a 100 mV threshold with respect to Pin 15. A logic 0 for this input is defined as < 85 mV, and a logic 1 is > 115 mV.

3. The fault and top drive outputs are open collector design and active in the low (0) state.

4. With 60°/$\overline{120°}$ select (Pin 22) in the high (1) state, configuration is for 60° sensor electrical phasing inputs. With Pin 22 in low (0) state, configuration is for 120° sensor electrical phasing inputs.

5. Valid 60° or 120° sensor combinations for corresponding valid top and bottom drive outputs.

6. Invalid sensor inputs with brake = 0; All top and bottom drives off, $\overline{\text{Fault}}$ low.

7. Invalid sensor inputs with brake = 1; All top drives off, all bottom drives on, $\overline{\text{Fault}}$ low.

8. Valid 60° or 120° sensor inputs with brake = 1; All top drives off, all bottom drives on, $\overline{\text{Fault}}$ high.

9. Valid sensor inputs with brake = 1 and enable = 0; All top drives off, all bottom drives on, $\overline{\text{Fault}}$ low.

10. Valid sensor inputs with brake = 0 and enable = 0; All top and bottom drives off, $\overline{\text{Fault}}$ low.

11. All bottom drives off, $\overline{\text{Fault}}$ low.

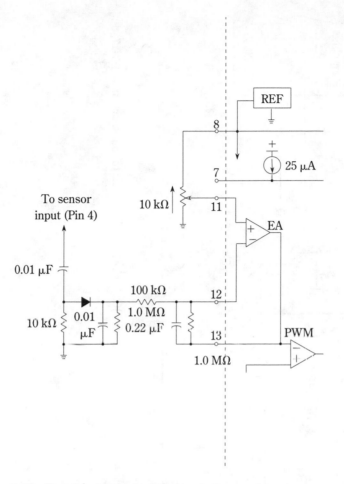

7-17   Closed-loop speed regulation via the use of a motor sensor. The pulses obtained from the HALL-effect sensor are proportional to motor speed. Therefore, a physical tachometer is not needed. Motorola Semiconductor Products, Inc.

Further insight into the operation of the MC33035 open-loop system can be gleaned from a study of the truth table of Fig. 7-19, and the waveform diagram of Fig. 7-19, in which two operational modes are depicted, one for full speed (100-percent duty cycle), and the other for reduced speed at 50-percent duty cycle. Even though only the bottom-drive outputs can be duty-cycle modulated, the three motor drive currents are symmetrical. Thus, for the reduced-speed operation, you see the PWM ripple on the A, B, and C phases of motor drive current.

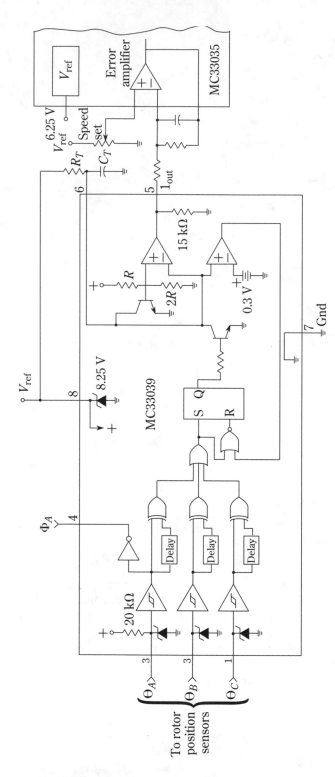

7-18 Closed-loop speed regulation via the use of the motor sensor and a dedicated IC. The principle is similar to that shown in Fig. 7-17, but the special MC33039 allows for more precise regulation. Motorola Semiconductor Products, Inc.

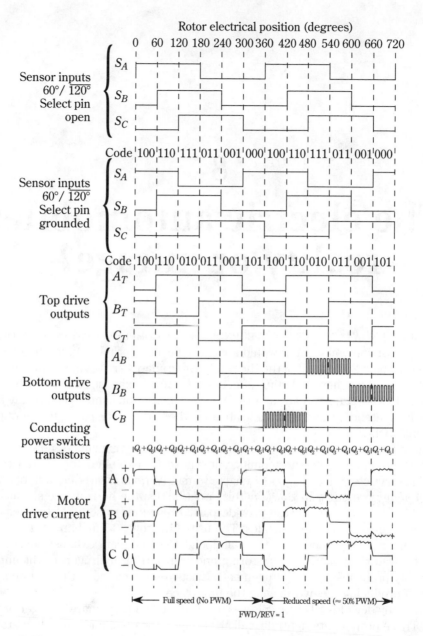

7-19 Three-phase, six-step, full-wave commutation waveforms. Full-wave infers bidirectional current in each stator winding. Motorola Semiconductor Products, Inc.

8

The electric automobile: reality or mirage?

IF WE DIDN'T KNOW OTHERWISE, THE BURNING OF THE COMPLEX SUBSTANCE GASOLINE, in internal-combustion engines would appear to be a bad propulsion method for automobiles. About 20 percent of the thermodynamic energy arrives at the tire road interface. The atmosphere polluting effects of the toxic exhaust products would be unthinkable were it not already an everyday fact of life. You cannot too quickly fault those who contend it was a mistake to abandon the horse as the prime mover of people and goods!

The dilemma of the gasoline-dependent auto is easily explained, however. For notwithstanding all of its shortcomings, it has so far exhibited the overall transportation characteristics of any fuel-engine format. Various environmental, political, and economic pressures might alter this situation. On the horizon, we see alcohol and methane-fueled cars. These provide partial remedies to some of the negative features of conventional automobiles. The use of hydrogen to fuel an internal-combustion engine looks good from the standpoint that the main exhaust constituent would be water vapor. However, storing and transferring hydrogen might impose some nasty problems. Also, hydrogen enthusiasts tend to overlook the cost and side effects of producing the hydrogen in the first place.

For many decades, electric vehicles have not lacked their share of eager proponents. These proponents rightfully call attention to the quiet and smooth operation of such cars. And, it is easy to support their contention that electric vehicles are simple and reliable, being devoid of the hundreds of moving parts (many of them reciprocating) in gasoline-motored autos. They also emphasize the lower order of maintenance probably needed.

However, these proponents tend to overlook some disturbing aspects of their euphoria. For example, it is true that the electric vehicle need not pollute the air from its own operation. But, the electric generating utility *will* be obliged to eject more pollutants into the atmosphere. Probably, in balance, fewer toxins will hover over cities, however. Also, a beneficial trade off could result considering the *nature* of the polluting substances on a global scale.

It is sometimes also overlooked that, thus far, electric vehicles don't appear likely to rival gasoline cars in performance. Any increase in power or acceleration is at the expense of *range*, which remains at short supply at best. Then too, the owner of the electric vehicle will probably have to allow for overnight charging from his residence. Special stations would be available for faster charging, but nothing approaching the two or three minutes it takes to refill a gasoline tank. Finally, this owner will probably have to cough up a sum for battery replacement at two to several-year intervals—all practical batteries deteriorate from repeated charge-discharge cycles.

Although dramatic breakthrough has not been made, the electric vehicle is steadily gaining a respectable measure of public acceptance, and is finding its own special niches. The laws of physics and the constraints of practical engineering do not stand in the way of continuing progress in batteries, motors, and control techniques. At some point, the more general market will decree overall performance to be worthy of investment. That time might be close at hand, especially if the cost factor can be relaxed via mass production—admittedly easier said than done.

The lead-acid battery: survivalist-hardened energy source

The archiles heel of the electric vehicle in 1913 was the battery. Even then, it was all too evident that the energy stored in a gallon of gasoline greatly exceeded the energy stored in either the same weight or volume of an electrochemical storage device. The practical manifestation of this was that automobiles with internal-combustion engines and fueled by gasoline could easily outperform electric vehicles despite the much lower overall efficiency of gasoline cars. Particularly distressing has been the limited *range* between charges in electric vehicles. The lead-acid storage battery was used in 1913 electric autos, and it is amazing that this type of battery remains a viable energy source for electric vehicles.

Battery technology is a voluminous subject. However, it is not necessary to cover the subject in great detail. Despite the investigation of almost countless battery types (many using exotic materials and sophisticated operational techniques), most prove deficient in *some* performance characteristic. The frustrating aspect is that there invariably appears to be unacceptable trade offs between important parameters. For example, the highly tauted sodium-sulfur battery must operate at an elevated temperature (about 570° F) to keep both, the sodium and the sulfur, in a molten state. The container for such a battery is like a thermos bottle in order to prevent appreciable heat leakage. Although such a battery outperforms the simpler lead-acid type, the public is not enthused by the prospect of being doused with either highly reactive molten sodium, or by hot liquid sulfur in an accident. Incidentally, advanced lead-acid batteries are of the gel type, in which relatively little sulfuric acid electrolyte is retained in sponge-like material.

A unique attribute of the venerable lead-acid battery is its ability to provide high levels of power for short durations. Indeed, that is a major reason it has been so useful in conventional automobiles where tremendous starting current might be demanded by a large engine on a cold morning. For the electric car, such capability is needed for passing, acceleration, and hill climbing. Interestingly, one overseas electric-vehicle uses exotic batteries to gain range from superior energy storage, but also carries enough lead-acid batteries to supply short-time demands of high power.

Many exotic batteries have aroused premature enthusiasm with high specific energy, acceptable specific power, manageable cost, and tolerance of wide ambient-temperature swings, only to fail miserably because of limited life span. All practical batteries deteriorate with charge-discharge cycles. Again, the lead-acid battery is found to provide a compelling balance among all performance parameters, including initial cost. Until exotic batteries can prove as impressive on the road as in the benign environment of the laboratory, the time-proven lead-acid type is not destined to be quickly obsoleted.

The chemical action in the lead-acid cell

The lead-acid battery has always been recognized as possessing a very desirable feature—it could be restored to use after being discharged by reversing the direction of current flow. In other words, such a battery could be recharged by receiving energy from an external current source. Such behavior stands in contrast to a large class of batteries in which discharge is a unilateral process; either one electrode is consumed, or nonreversable chemical changes render such batteries permanently unsuitable for further use. Thus, we have the two major types of cells or batteries—the primary batteries that are good for a *single* discharge action, and the secondary or storage batteries that can be restored to active performance by *recharging* from an external current source.

The chemical action within the lead-acid cell is depicted in Fig. 8-1. Initially, one electrode consists of lead dioxide, PbO_2. The other electrode is lead, Pb. These are immersed in a sulfuric acid electrolyte, H_2SO. The electrolyte provides dissociated positive hydrogen ions, H^+, and negative sulfate ions, SO_4, as current carriers *within* the cell. These ions charge the electrode plates, which promote electron flow in the *external* circuit. Ultimately both electrodes become coated with lead sulfate, $PbSO_4$, and we no longer have an active electromotive cell. In the process of discharge, water is formed that dilutes the electrolyte. That is why a hydrometer can be used to sample the density of the electrolyte, and to infer from it the state of charge of the cell. The nice thing about the chemical reactions is that they can be *reversed* by reversing the current flow for an appropriate period of time.

This behavior is fortuitous in terms of use in electric vehicles. The owner does not need to handle dangerous chemicals, or be inconvenienced by requiring services from a specialist, or a specialized facility. He or she only has to operate a battery charger overnight from his utility wall outlet (this, admittedly, looms up as an inconvenience and much effort will be directed toward the shortening of this period).

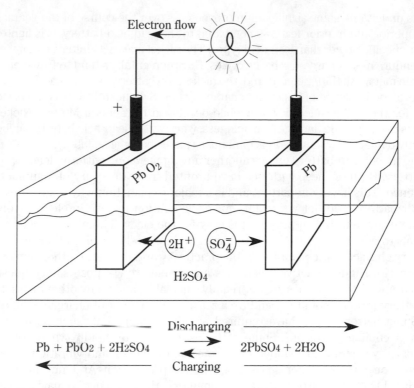

8-1 Chemistry of the lead-acid cell. As discharge commences, both plates become converted to lead sulfate, $PbSO_4$. At the same time, part of the electrolyte becomes converted into water, H_2O. The *reverse* reactions occur when the cell is charged from an external source of dc.

Even the relatively dilute sulfuric acid is a much less harmful agent than the chemical substances used in some of the more exotic battery types. Replenishment of the acid is not necessary; only distilled water need be added to compensate losses from evaporation and electrolysis. Even this is not necessary in advanced designs. A further improvement is the use of a gel instead of liquid electrolyte. And finally, lead is an inexpensive and benign material. It is easy to see why research programs not only seek out new exotic-battery formats, but focus considerable effort in further improvements of the venerable lead-acid battery. Its specific energy of about 40 or possibly 50-Wh/kg is not impressive, but its overall features adapt it well to electric-vehicle service.

The nickel-iron battery: another enduring tough one

The lead-acid battery has a companion type that was also in use during the turn of the nineteenth century. It was the nickel-iron battery with potassium hydroxide electrolyte as devised by Thomas Edison. It, too, possesses some notable qualities and has undergone some significant improvements over the course of the years. It has both superior and inferior features compared to the lead-acid battery. It is lighter, but has less specific power than lead-acid types. The nickel makes its initial cost higher, but it can endure more charge-discharge cycles. Environmentalists tend to frown on nickel as a toxic metal. All things considered, the nickel-iron battery, like the lead-acid battery, exhibits a good balance of important characteristics. Significantly, it was the battery of choice for the Impact electric-automobile developed by General Motors. Nickel-iron batteries undergo little damage from idleness, whereas the lead-acid battery can be permanently damaged by long periods of nonuse (plates become sulfated).

The nickel-iron battery enjoys another advantage along with the lead-acid storage battery. Both of these batteries have profited from a century of manufacturing experience. This is very important and is a major explanation why many apparently superior batteries do not yet qualify for consideration for use in electric vehicles. From a purely textbook approach, it is easy for the electrochemist to select battery materials to maximize cell voltage, minimize weight, and optimize other parameters of battery characteristics. For use in the practical world, however, other things enter the scenario. Highly reactive electrodes and/or electrolyte pose safety questions. The charging characteristics are not readily calculable, and many otherwise satisfactory battery formats could not endure a reasonable manner of charge/discharge cycles. Some battery types have required auxiliary equipment, such as electrolyte pumps or cooling provisions. Some batteries lose charge rapidly when not in use. Some battery systems are vulnerable to mechanical shock and vibration. Many are too costly, or don't lend themselves to easy mass production. And, not infrequently, unpredictable "bugs" turn up. A case in point was the "memory" behavior that long plagued the operation of nickel-cadmium batteries.

Summarizing, the lead-acid battery and the nickel-iron battery remain viable energy sources for electric vehicles, and will remain so until displaced by higher performance types that can also comply with the many practical demands of the road.

Basic chemistry of the nickel-iron cell

Although different materials are used in the Edison, or nickel-iron battery, it shares important features with the lead-acid battery. Most important, both formats qualify as secondary batteries because they permit replenishment of stored electrochemical energy by the simple technique of recharging. Instead of an acid electrolyte, the nickel-iron cell makes use of potassium hydroxide; for this reason, it is sometimes referred to as an alkaline cell. Interestingly, the density and composition of the electrolyte remains fairly constant regardless of the state of charge. Thus, the use of hydrometer is not indicated for ascertaining the charge condition.

Some unique features of this battery include: it is not very susceptible to damage from overcharging, it remains functional in colder climates than does the lead-acid battery, it can be stored in inactive condition for many years with negligible chemical deterioration, and it is a lighter battery than the lead-acid type. Obviously, if the vehicle can haul less dead weight, its *range* can be extended. A shortcoming is that the nickel-iron cell develops about 1.4 volts compared to the approximately 2.15 volts of the lead-acid cell. Thus, a designer must incorporate more cells in order to attain a desired operating voltage in the vehicle.

When a nickel-iron cell is charged, its positive electrode is a higher oxide of nickel and its negative electrode is iron. During discharge, the positive electrode is converted to a lower oxide of nickel, and the negative electrode becomes iron-oxide. Recharging from an external source of current reverses this change in electrode composition. In essence, you have the transit of oxygen ions in one direction or the other depending on whether the bell is being discharged or charged. A chemist would say that the iron electrode (negative terminal) undergoes oxidation during discharge, and that the nickel-oxide electrode undergoes reduction during discharge. Oxidation and reduction then take place at opposite electrodes when the cell is charged. The nickel-oxide electrode alternates between the higher oxide of nickel, NiO_2 and its lower oxide, NiO.

An unusual constructional feature of the iron-nickel cell is that its container and its various mechanical support elements are largely made of steel. No corrosion, undesired chemical reactions, or interferences with the cell operation are experienced from this fabrication technique, which actually contributes to the long life and stability of this cell format. Where necessary, hard rubber, or more advanced insulating materials are used for separators between plates, and where conductive supports are ruled out.

Batteries—Plain and fancy

It is not easy to define a battery format that is "exotic". In this book, the general idea will be that batteries other than lead-acid or nickel-iron types qualify as exotics, at least from the standpoint of applicability to electric vehicles. Admittedly, some battery types will occupy nebulous positions in which a valid argument can be presented regarding their exotic status. For example, it is easy enough to question the "exoticness" of nickel-cadmium batteries, which have had many years of experience in all manners of practical power applications. However, the very fact that this battery format has, and continues to evoke, strong objections for vehicular use because of environmental, cost, marginal performance, safety (real or perceived), or other reasons inhibits wide acceptance for powering electric vehicles. Because a battery format is labeled exotic doesn't mean that it has no potential for consideration. Indeed, the very opposite tends to be true; they generally possess a cluster of very compelling attributes in conjunction with one or a few shortcomings. Exotic batteries are worthy of mention for the simple reason that much time, money, and effort is invested in endeavoring to overcome the shortcomings. Most certainly, some of these efforts will meet with enough success to merit consideration.

Keep in mind that the near-term objective of improved battery performance is not to necessarily match or exceed the performance of conventional automobiles. However, there is general consensus that the *range* between recharging should be on the order of at least two-hundred miles of "average" driving conditions. The cost should not be exotic—this takes into account the number of replacements needed over, say a five- or six-year term of ownership. Although "quick charging" cannot be done from the residence wall outlet, the battery should be amenable to shorter than the generally prescribed eight-hour period. Although not an impenetrable barrier, the need for elevated temperatures and the reliance on highly reactive agents are negative factors. Very superior performance would be necessary to overcome these negative factors. (That is why the public accepted the conventional automobile even though it uses the highly volatile, flammable, and potentially dangerous fuel, gasoline). Some of the exotic electrochemical systems that have at one time or another been tauted as the "breakthrough" for the electric vehicle, or that could conceivably show favorable prospects in the future are listed in Table 8-1.

Table 8-1. Some exotic battery systems with possible application to electric vehicles. Watch for careless comparisons between kilowatt-hours per *kilogram* and kilowatt-hours per *pound* in the technical literature. One-kilogram = 2.205-pounds and one-pound = 0.4536-kilogram.

Batter format	Features	Shortcomings	Comments
Zinc/Nickel-Oxide	Small volume High specific-energy, approx. 100-Wh/Kg	High cost Poor charge-recycling life	Researched extensively by General Motors. Acceptable balance of parameters not reached
Zinc/Air	High specific energy, approx. 100-Wh/Kg	Low specific-power Recharged difficulties (Replenishment of reaction material by mechanical substitution is one "charge" technique.)	Some similarity to a fuel cell in that oxygen is consumed from the air. Basic format was once classified as a primary battery (Single-discharge behavior)
Zinc/Chlorine	High specific energy—about 150-Wh/Kg	Safety factor from chlorine gas. Mechanical pumping and refrigeration are involved.	Example of system which looks good in laboratory tests, but is difficult to adapt to the rigors of use in electric cars.
Aluminum/Air and Lithium/Air	Very high specific-energy, on the order of 7000 Wh/Kg for both.	Reliability problems and operational "bugs" not easy to overcome.	Fuel cells obviously remain in the race for the energy source of electric vehicles. In their present developmental state, they appear unsuited for use by the average motorist.
Sodium/Sulfur	High specific-energy, on order of 200-Wh/Kg "Solid" electrolyte	High temperature operation at 570°F. Dangerous reagents in case of accident.	Ford has developed this format as a viable energy source for electric vehicles. Public response remains to be seen.
Lithium/Metal-sulfide	High specific-energy coupled with high specific power.	Operates at 750°F. High cost of boron-nitride separator material	Needs more work. Advocates cite available waste-heat for vehicle heating and A/C.
Lithium/Polymer	100-Wh/Kg "Solid-State"—no gel or liquid electrolyte. Easily manufactured in variety of shapes. Low production costs.	Improved low-temperature performance is needed. Environmentalists do not look favorably on lithium.	General Motor's Delco division is attempting to develop this battery format to replace conventional lead-acid batteries in gasoline-powered cars. If successful, technique could be further pursued for use in electric vehicles.

"Fill it up" . . . but, how quickly?

When evaluating the characteristics of a battery system for an electric vehicle, it is almost instinctive to hope for the ability to recharge quickly. In such a quest, however, it is all too easy to overlook a practical barrier to rapid charging, other than the battery itself. Consider, for example that a vehicle has a battery supply rated at 60-kilowatt hours. Suppose that half of this energy has been used up and it is desired to replenish the fully charged condition from the 120-volt wall outlet of a typical residence. To accomplish this in ten hours, the required current drain from the household would be:

$$\frac{30 \text{ kw-hrs}}{120 \text{ V} \times 10 \text{ hrs}}$$

$\frac{1}{40}$ kilo-ampere, or 25 amperes. So far, so good. Certain electrical appliances do successfully run on 25 amperes in the home. For simplicity, the less than 100 percent efficiencies of the rectifier and the electrochemical process within the battery are ignored in this example. The gist of the idea being developed will not be materially affected, however.

In the example, as so far presented, the charging technique is practical, even if it be argued that it is only marginally so. But now, suppose that you are somewhat less patient and demand replenishment of your "fuel" in one hour. The wall outlet would then be called upon to supply 250 amperes under the postulated conditions. This would be clearly unacceptable and in a practical sense would constitute a short circuit.

What might help? Many homes have 240-V optional outlets. The use of this option might reduce the required charging time to five, or possible four hours. For quicker charging than this, it would be necessary to go to a special "filling station" where the time could be further reduced, say to an hour or less. It is not too easy to visualize much faster service, although it is not clear exactly where to establish a limit to such rapid charging. Such a limit might exist because the battery itself might have its life threatened or shortened by an excessive charging rate. The mindset of the owner of an electric vehicle will have to change to comply with the new circumstances—no longer is he likely to enjoy the luxury of refilling the "tank" in a couple of minutes or so.

This problem might bring into being battery or fuel-cell systems in which spent electrodes and/or electrolytes are physically replaced by new materials. This might conceivably be done in some minutes, but one would hope there would be no detrimental trade off in other performance features. Then, there is the possibility of temporarily trading one's discharged battery for a quickly replaceable fully charged one. The economic and mechanical practicability of this strategem is open to question, however.

No rest for the virtuous

There is another aspect of the recharging scenario that tends to hold back the popularization of electric vehicles. The matter of charge rate has already been alluded to and it will be assumed that the public is willing to live with appreciably slower energy replenishment than prevails for conventional automobiles in which a quick visit to the gas station suffices. This has to do with the nature of the hardware that must be handled by the motorist to recharge his vehicle battery; it merits special attention because it involves *safety*. It looms up as a particularly nasty obstacle, but you might be heartened by the fact that the public routinely handles the gasoline pump—a potentially dangerous operation that we just take for granted because it has been in place for so many years. We have learned refrain from smoking while refueling, but otherwise we don't give much thought to the flammability of the volatile liquid we are dispensing.

In the case of the electric vehicle, there is the matter of electric shock from the 120-V ac line, or worse, from 240-V lines that will be used to gain shorter recharges. Recessed connectors and other hardware innovations can be used, but with millions of electric vehicles (eventually) being recharged around the clock, the probability for injurious or lethal electric shock is not negligible. The danger would be enhanced in wet weather, and would increase with extended use of the cables, plugs, and receptacles. The problem is also related to the controversial issue as to whether the charging equipment and/or cables should be carried on board the vehicle, or should they be incorporated in a stationary location. The vehicle manufacturer is reluctant to carry the charger in the vehicle because it counters the objective of transporting as little weight as possible. A compromise might readily be made, however, if the charger is not too massive; this implies that the charging rate cannot be too high.

Another approach—one likely to receive serious consideration—uses inductive coupling to transfer energy from the ac utility line to the on-board battery charger. The owner would merely have to position his or her vehicle over an induction loop on the floor. Because of limited ferromagnetic material, the efficiency of this technique would best be served via frequencies in the supersonic, or low RF region.

As with many of the exotic batteries, here is a disparity between what the technologist knows is feasible, and the mundane requirements imposed by practical ownership of electric vehicles. Commendable performance and affordability are not enough if the public suspects problems of safety, reliability, or undue convenience. A little contemplation shows that these matters are closely interlinked.

Can the utilities handle the load imposed by electric vehicles?

The arrival of popularity of electric vehicles will not be an isolated phenomenon. A supporting infrastructure will be necessary. Some of the elements are already in place, such as roads, traffic signals, and utility power. However, the latter element requires some scrutiny, because it is the most important—it is the utility that really provides the energy for the electric vehicle, even though storage batteries are used as intermediary agents. The all-important question is whether electric vehicles will adversely impact the electrical generating station. This will decide whether electric vehicles are destined to fill a meaningful transportation niche, or whether they will continue as interesting toys of technically inclined hobbyists.

We might conceivably be pessimistic regarding this matter, for it is common knowledge that utilities have been complaining about *overloads*. In the summer, air conditioners sap a great deal of energy. In the winter, electric heaters tend to be the culprit. It might seem that the additional load imposed by battery chargers replenishing energy-hungry batteries could be the straw that breaks the camel's back. Massive upsizing of generating capacity would surely run into economic roadblocks; also, delays would likely be brought about by environmentalists. Thinking in these terms, it appears that our discussions of batteries, motors, and control techniques is an interesting academic exercise, but devoid of practical relevance.

Fortunately, it turns out that advocates of electric vehicles have both brains and luck. The utility companies would actually *welcome* such additional loading, although under certain easily complied with conditions. The size of a generating station is determined by the *peak load* it expects to handle. But, during off-peak hours (during night time) the larger than necessary facilities are expensive to run—efficiency is low and there is not much revenue. For this reason, some means would be imposed to induce electric vehicle owners to do their charging at night, a not necessarily inconvenient activity. Calculations reveal that it will be many years before most utilities will need to increase their energy capacity. It will be important in many cases, however, to discourage too much daytime recharging.

It has already been pointed out that although large-scale use of electric vehicles would shift the pollution source from city streets to the remotely located utility station, a worthwhile reduction in *net* pollution would nevertheless result. Not only would there be less overall pollution, but the solution products indirectly attributed to the energy demand of electric vehicles would be relatively benign compared to the exhaust gasses from the internal-combustion engines of conventional automobiles.

dc or ac motors?

For many decades, the dc series motor had been the basic workhorse of various types of electric vehicles. Sometimes, an extra shunt field was incorporated and the compound motor thereby created enabled better optimization of the speed-torque relationship, as well as refinement of control. Later, when high-flux permanent-magnet motors became practical, these motors were also found to be satisfactory. And when solid-state power devices began to be used as choppers or duty-cycle modulated control elements, it was felt that the drive/control functions had at last attained mature technological status in the electric vehicle.

However, these same solid-state devices also enable design and construction of very efficient *inverters*, and it wasn't long before serious consideration was given to the use of ac motors. This was largely inspired by the desire to eliminate the commutator and brushes inherent in conventional dc motors. The attainment of this objective would dispense with a major maintenance item. Enthusiasm was enhanced by the fact that these solid-state inverters could be implemented in two-phase and three-phase formats, and the frequency could be readily controlled. Successful drive power was achieved with induction motors, and also with synchronous motors. Additionally, so-called brushless dc motors were demonstrated as successors to the venerable dc series motor. Giant stepping motors offered yet another way to circumvent the need for brush and commutator motors.

Actually, an ongoing controversy developed between the proponents of the time-proven dc series motor, and those who see ac motors as harbingers of advanced technology. Both design philosophies are imbued with much merit. Despite the mechanical motion and sliding contact of brush and commutator motors, an electric vehicle appears able to accumulate about 80,000 miles before brush and commutator "tune ups" are needed. And compared to various repairs required in an internal-combustion engine, such maintenance is not major surgery. Also, it must be conceded that when such a motor is duty-cycle controlled, the overall performance is very satisfactory.

On the other hand, even better efficiency can be realized with ac motors. They tend to be smaller and lighter. Another advantage is that they lend themselves better to total enclosure, thereby keeping out dirt and grit. Polyphase induction motors and synchronous motors are less costly to manufacture. But it is difficult to find overwhelming evidence in favor of either type of motor. In the near term, we might expect to see *both* types used in electric vehicles. This is especially true because electronic control enables desirable performance characteristics to be tailored into a variety of motors no matter what the natural behavior of the motor might be.

Reducing losses with germanium power transistor choppers

An electric golf cart is a relatively primitive electric vehicle, somewhat reminiscent of "horseless carriage" automobiles marketed during the early years of this century. In particular, the use of the dc series motor and lead-acid storage batteries can cause you to ponder the progress of technological development made three-quarters of a century or so ago. There are, of course, more sophisticated golf cart designs, some with 10-horsepower ac motors, but the described format still yields satisfactory performance.

The dc motor is usually the series type, often with a dual series winding arranged to allow convenient reversal of motor rotation. Some designs have used permanent-magnet motors and compound motors. In any event, dc motors of about two-horsepower capacity seem to provide optimum compromise among performance and cost factors. The storage batteries are similar to those used in automobiles, except that they are designed for long survival in the face of many deep discharge/charge cycles.

The control of speed by variable resistance, although workable, would be undesirable because of the waste of energy and the shortening of serviceable time between battery charging. A dc chopping technique is much better. Indeed, the ability to implement such control is one of the greatest advantages over the early electric automobiles. A simple and effective chopping format can be provided by the PWM control-ICs commonly used for switching-type power supplies. The modulation can be either constant frequency, variable pulse width, or constant pulse width, variable frequency. In either case, the average voltage applied to the motor is varied by a means that would be 100 percent efficient in ideal form.

For golf carts, at least, germanium transistors merit consideration as the power switches to supply the chopped dc to the motor. Germanium power transistors are inexpensive and exhibit very low collector saturation voltages, making dissipation from conductive losses low. However, successful use of those devices often requires ways to deal with their switching losses and their tendency towards high leakage current. The partial circuit of an electric golf cart shown in Fig. 8-2 illustrates practical methods of circumventing much of these losses, while still benefiting from the desirable features of germanium power transistors. Notice the two departures from conventional circuits of this variety: All germanium transistors have turn-off biases applied to their bases and there is an SCR connected across the paralleled bank of germanium switching transistors (Q3 through Q8).

The consequence of this arrangement is that the part of the switching loss inherent in the slow turn on of the germanium transistors is overcome; the SCR turns on much more quickly. Moreover, the motor current is carried by the transistors for *most* of the conduction time. This comes about because of the lower voltage drop in these transistors. Thus, each device is allowed to be active where it is advantageous. SCRs alone would cause higher conductive losses; germanium transistors alone would cause higher switching losses. The nice thing about this hybrid scheme is that the desired transitions from one type of switching device to the other occur automatically.

It is natural enough to ponder the relevance of discussing a golf cart when our investigation of electric vehicles purports to deal with a possible substitute for the gasoline-powered automobile. Many electric car enthusiasts have often used scaling-up techniques on golf cart designs to produce an electric automobile. Also, to satisfy the needs of the golf cart, battery manufacturers have developed lead-acid batteries that feature good longevity in the face of many repeated deep-cycle discharges. Thus, these batteries proved tailor made for many experimental electric autos. The inordinately high electrical ruggedness of these golf cart batteries also enables *faster* charging rates. This sets well with electric car owners, whose patience for this interlude is notoriously in short supply.

Germanium power transistors are now made by Germanium Power Devices Corporation of Andover, Massachusetts. This firm has taken over the device lines of Motorola, RCA, G.E., and others, and has developed its own family of improved devices. Readily available are 25, 50, and 100 ampere power transistors. When experimenting with germanium transistors, keep in mind their polarity requirements inasmuch as they are PNP devices. Fortunately, in vehicle applications, massive heat sinking is usually not objectional and it is not difficult to operate germanium transistors at moderate temperature rises. This alone circumvents one of their common shortcomings in nonvehicular uses.

Experimenters with electric vehicles should also be aware that this company has developed germanium Schottky diodes with several hundred ampere capability. These drop less voltage and waste less power than either silicon junction diodes or silicon Schottky rectifiers. A possible use might be in battery systems combining two different types of batteries, usually one for high specific *energy* to give the car a long range, the other for high specific *power* to enable sudden acceleration, and to aid in hill climbing. Electrical isolation of the two battery types is achieved by connecting them in parallel through a diode. Each battery is then free to undergo its own independent charging rate.

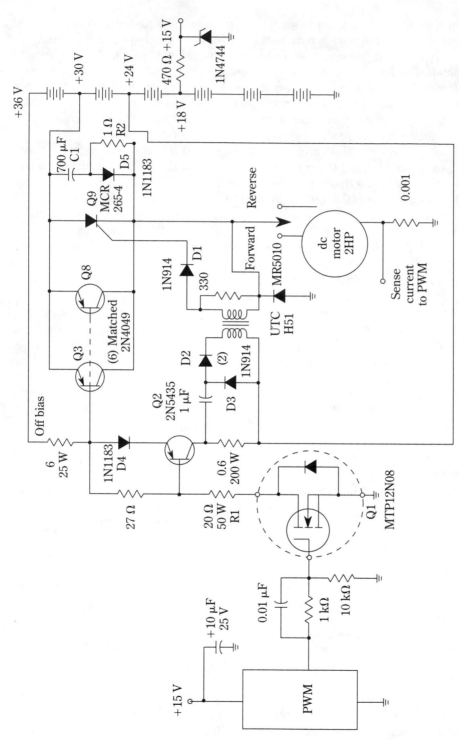

8-2 Partial schematic of efficient golf cart motor-control system. Dissipation from leakage-current of the germanium transistors is reduced by turn-off bias applied to their bases. Also, their slow turn-on is overridden by the SCR (Q9). During conduction time, the motor current is carried by the bank of parallel-connected germanium transistors (Q3 through Q8). *Motorola Semiconductor Products, Inc.*

Conversions:
A practical path to electric propulsion

It is somewhat of a paradox that the "most bang for the buck" electric cars have been designed, produced, and marketed by small-scale entrepreneurs. For a much higher initial cost, the large corporations have developed electrics with exotic batteries, elegant motor-drive systems, and sophisticated electronic controls. But, thus far, they have not impressed the public with their claims of performance, reliability, convenience, and overall ownership expenses. The situation is not unlike the auto market during the great depression when many different makes of cars competed for the consumer's interest by offering a myriad of then state-of-the-art innovations. It turned out that the best-seller was the economy-priced, no-nonsense Model A Ford.

With electric cars, too, the public has been best served with straightforward, time-proven systems involving low-cash outlays and capable of readily acquired maintenance. This generally translates into the *conversion* of light four-cylinder automobiles to electric propulsion. Most often, a dc series motor fed from heavy-duty lead-acid batteries via a variable duty-cycle electronic chopper comprises the "guts" of the installation. Cleverly designed kits are offered that enable the purchaser to easily replace the gasoline engine with the dc motor. Often, no welding, machining, or special mechanical procedures are involved in the conversion. Typical performance ratings are top speeds of 65 mph and ranges of 60 to 80 miles. Advertising emphasizes practicability within the limited performance capability—the public is not deceived to expect comparable performance to gasoline cars.

An example of such electric cars is shown in the photograph of Fig. 8-3. From the exterior, you see what appears to be a "normal" VW Rabbit four-passenger sedan. However, as shown in Fig. 8-4 the transverses internal-combustion engine and its peripherals have been removed. In its place, are the lead-acid storage batteries, but the dc motor is beneath them, not visible (the cylindrical object is the vacuum tank for the brakes). Figure 8-5 shows an installation. The eight-inch diameter dc motor is shown mounted to the car's transmission in Fig. 8-6.

The conversion kit is marketed by Electro-Automotive, P.O. Box 1113, Felton, California, 95018. All hardware, cabling, adaptors, controls, as well as special tools and heavy-duty springs and shocks are included. User-friendly instruction are also provided—the author interviewed a young lady who made the conversion with minimal outside assistance and was very pleased with the smooth, noise-free operation of the car. The overall cost of the kit was about $7,000 in 1992. A nice feature of this electric-car conversion is that the charger is carried aboard, enabling the driver to place the batteries on charge anywhere a 60-Hz, 120-V outlet is available.

From the user's vantage point, an electric car of this type has some compelling advantages when compared with more exotic designs. To begin with, the initial cost tends to be half or less of ordinarily encountered outlays. Note that the original transmission is retained. This confers the psychological benefit that it is "natural" to drive such an electric car (it has been determined that the loss of efficiency from retention of the gear box is not significant). The transmission also allows reverse, without reversing the motor's rotation.

8-3   The VW Rabbit electric car. Externally, the converted four-door sedan retains its original appearance. Other light automobiles such as the Simca, Renault, and Hillman have been favorites for electric conversion. They tend to be inexpensive on the used car market, but it is generally necessary to beef up their suspension systems to accommodate the weight of the batteries. Electro Automotive—Felton, California.

8-4   Engine compartment view of the electrified car. The dc series-motor cannot be seen because of its location beneath the batteries. The original transverse-mounted internal-combustion engine has been removed. Electro Automotive—Felton, California.

8-5 Installation of the dc series-motor in the engine compartment. The original transmission is fitted to the right-end of the motor. As with the original internal-combustion engine, transverse drive is used. Electro Automotive—Felton, California.

8-6 View of the dc series-motor mounted to the car's transmission. Actually, the purchaser of the conversion kit has optional selection of several similar motors with slightly different ratings. This enables optimization of range, speed, or acceleration in the converted vehicle. Electro Automotive—Felton, California.

Lights and other electric peripherals that operate from a single 12-V battery in conventional automobiles also are supplied from their individual battery in the electric conversion. This battery is separate from the bank of 8 batteries that power the dc series motor. Instead of an alternator, a dc-dc converter samples the nominal 96-V "line" and delivers approximately 13.5 V to this additional battery in order to maintain its charge. The purpose of this technique is to provide equal drain to all eight of the "main" batteries.

Although, the performance parameters admittedly fall short of those attainable from internal-combustion engined automobiles, the practical fact of auto transportation is that about 60 percent of motorists probably demand less than a daily 45-mile excursion to get to and from their jobs, and for shopping and miscellaneous purposes. It appears that an electric car of this type can fill such a niche and be found rewarding because of its reliability, relatively low maintenance, smooth operation, and its environmental cleanliness.

The smoothest path to an electric car via conversion is to first find out what conversion kits are available and then see about the availability of the appropriate vehicle. The Ford Escort happens to be a relatively easy car to deal with. In any event, what is needed is a light automobile with stick shift. The more mechanically adventurous often forego the kit and do the job from scratch. Popular motor experimenters have had success with the dc generators from large military aircraft. These have been available at government auctions and are often advertised by surplus outlets in hobbyist magazines.

The fuel cell

Most workers in the electric field agree that electric motors and motor-control techniques have forged ahead of the quest for a suitable source of on-board energy. As pointed out, the fancy numbers defining the performance of many tauted battery formats simply do not translate readily into practical implementations. Exotic batteries look good in the laboratory, but exhibit serious shortcomings for actual use. There is one electrochemical cell, however, that deserves special mention, if for no other reason that it appears to offer the greatest promise together with tantalizing disappointments.

I am speaking of the fuel cell. To those familiar with the chemistry of primary and secondary voltaic cells, and with other electrochemical systems, such as electroplating, the fuel cell does not appear as a radical departure from its electrochemical cousins. It is, however, a unique device. Unlike primary cells, neither electrode is consumed. And unlike secondary cells, neither electrode undergoes chemical change; moreover, the fuel cell cannot be recharged, nor does it need such energy replenishment. Rather, as its name implies, it operates from a fuel, the simplest of which is hydrogen gas. It must also be provided with oxygen, but this need is satisfied by admitting air. A nice thing about a fuel cell, fed these ingredients is that the exhaust comprises hot water and the unused nitrogen of the air.

A basic fuel cell is shown in Fig. 8-7. The anode and cathode electrodes are porous carbon surfaced with a catalyst, such as platinum, which also is porous, or is applied so as not to clog the pores of the electrodes. These electrode assemblies are separated by an electrolyte to provide ion conduction. A commonly used electrolyte

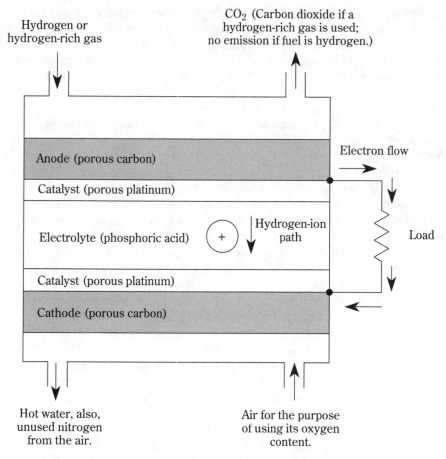

Hydrogen or
hydrogen-rich gas

CO_2 (Carbon dioxide if a
hydrogen-rich gas is used;
no emission if fuel is hydrogen.)

Anode (porous carbon)

Electron flow

Catalyst (porous platinum)

Electrolyte (phosphoric acid)   +   Hydrogen-ion
path

Load

Catalyst (porous platinum)

Cathode (porous carbon)

Hot water, also,
unused nitrogen
from the air.

Air for the purpose
of using its oxygen
content.

8-7 Basic structure of a fuel cell. Unlike batteries, the electrodes are neither altered nor consumed. The fuel cell has no need for energy replenishment by
charging. As long as hydrogen and oxygen are available at the electrodes, electricity is delivered to the load.

is phosphoric acid, which takes no part in any chemical reactions. Because both electrodes are of essentially similar materials, the device delivers no current as hitherto
described.

If, however, hydrogen or a hydrogen-rich gas is introduced into the anode region
and oxygen-bearing air is allowed to enter the cathode region, things begin to happen. Electricity is then supplied to an external load circuit. Let us see how this
comes about.

Hydrogen molecules consist of pairs of positively charged hydrogen atoms, or ions.
Ordinarily, there is a certain amount of separating and reuniting of hydrogen ions. In the
presence of a catalyst, such as platinum, the separating of the ions from the molecule is
greatly speeded up. As a result, many of these liberated ions migrate through the electrolyte and are attracted to the cathode, which bears a negative charge. But, when the
catalytic action at the anode splits the hydrogen molecule into its constituent ions, it simultaneously liberates two *electrons*, which also reside in the hydrogen molecule.

It is these *freed* electrons that pass through the external load and back to the cathode, giving it its negative charge. The electron transit through the load constitutes the electric current in the load. An interesting thing then happens in the cathode region. Here, the atmospheric-borne oxygen is dissociated into negative oxygen ions, again by catalytic action. It so happens that if you have oxygen ions, hydrogen ions, and electrons in one region, the easiest recombination for these charged entities to enter into is the formation of H_2O, or water.

From the foregoing, a complete cycle of events has been traced—hydrogen molecules are dissociated into charged hydrogen atoms, that is positive ions, in the anode region. This ionization process also liberates free *electrons* that provide the electric current for the external load, and then return to the cathode. In the cathode region the cycle is completed when oxygen obtained from the air is catalytically ionized, and exhaust water is formed from hydrogen and oxygen ions and electrons.

The exhaust water carries away some of the heat from processes that are less than 100 percent efficient. Also, the atmospheric nitrogen, serving no chemical purpose, is ejected from the fuel cell. In the event a hydrogen-rich gas, such as methane, is used instead of pure hydrogen gas as fuel, carbon dioxide will also be exhausted from the fuel cell.

Practical fuel cells of this type provide about three-quarters of a volt per cell. Thus, such a dc power source must consist of many series-connected cells to develop suitable voltage levels for electric vehicles, generally in the 75- to 300-V region.

This energy technology compels consideration from the standpoint of allowing acceptable *ranges* for the electric vehicle; the fuel cell's specific energy can greatly exceed that attainable from electrically charged batteries. However, there are unsettled controversies regarding the nature and cost of the hydrogen-rich fuel. Also, our simplified discussion doesn't deal with various problems occurring with the catalysts and the electrolyte in actual applications. Much will be learned from the recent experimental installations of fuel cells in utility stations, where they will provide standby and peak-load power. Also, fuel cells have given good accounts of themselves in spacecraft. It is interesting to observe that some of the high-performance exotic batteries seem to use "semi" fuel cells inasmuch as they, too, consume oxygen from the air.

A common sense appraisal of electric autos

Caution is the name of the game in attempting to justify or condemn the electric automobile. Numbers can be cited to advance arguments, but the true meaningfulness of numerical comparisons is often less than crystal clear. "X" battery might have three times the energy capacity of "Y" battery, but does such a statement take into account initial and maintenance cost, safety, reliability, environmental impact of its materials, mechanical ruggedness, charge retention, and recharge characteristics? Even more nebulous are the time-worn comparisons periodically made between gasoline-fueled cars and electric vehicles. It turns out that the energy problems faced by an industrial society are very complex; in the larger picture, the usual assertions about a preferred transportation mode often loses relevancy.

Consider oil in the ground of an overseas country. That oil must be extracted and transported at costs that are as likely to be influenced by politics, intrigue, and manipulation, as by economic forces alone. A barrel of oil represents a quantity of energy.

Energy is lost when the oil is refined to gasoline. Energy is lost when the oil is burned in a fossil-fuel utility station. Energy is lost when electricity is sent to the customer's wall outlet. Energy is lost when gasoline is converted to thermodynamic energy, thence to mechanical energy in the internal-combustion engine and transmission of conventional automobiles. Energy is lost in the discharging and recharging of electric-vehicle batteries. All of these energy losses beget various cost and environmental consequences. The relative importance of these consequences depend in the first place upon the mix of fossil fuel, hydro, and nuclear generating stations. Also much involved are the demographic features, life-styles, and infrastructures of the society. It should come as no surprise that many assertions about electric vehicles vs. gasoline autos essentially compare apples and oranges.

If one attempts a consensus of electric vehicles predictions, a list of generalized statements can be compiled. Of course, all bets would be off in the event of a dramatic technological breakthrough on either side. A battery or fuel cell clearly able to extend presently practical ranges by a factor of five could speedily usher in the electric vehicle era. By the same token, a combustion technique extracting double the mechanical energy at one quarter the toxic emissions from the gasoline engine would likely ensure retention of this locomotion mode. In any event, the generalized statements that presently appear valid are as follows:

- The public wants more than performance suitable for nearby shopping excursions. Payload, acceleration, speed, cost, and range should not lag far from that attainable in a small gasoline-fueled economy car.

- Toxic fumes, such as carbon monoxide, carbon dioxide, nitrogen oxide, and ozone will be much reduced in urban areas. Sulfur dioxide will tend to be increased in the fossil-fueled generating station, but it will be much easier to cope with this than to suppress emissions from millions of automobiles.

- Despite fancy numbers attributed to a host of exotic chemical formats, the first batteries are likely to be advanced lead-acid, nickel-iron, and in more limited number, nickel-cadmium types.

- These rather mundane batteries will eventually give place to more sophisticated types involving more reactive electrodes such as lithium or zinc. Some of these will be air breathing, and will otherwise resemble fuel cells. Polymer and other solid electrolytes will merit serious consideration, but preference will be given to those formats that can operate at ambient, rather than elevated temperatures.

- Both dc and ac motors will probably be used. However, dc motors could gradually shed their brushes and commutators. As time goes on, there might be less difference between dc and ac motors because *both* will be heavily dependent on the logic of control electronics.

- Many small companies will likely get into the act, but domination will be attempted by the "big-three" auto makers. Small firms will have to be nimble and innovative to qualify for a slice of profits from the electric vehicle pie. Advertising hype will be dangerous to both small and large manufacturers—claims of extend range that pertain to unrealistic traffic conditions will damp the public's enthusiasm for electric cars.

- The smoothness, quietness, and maintenance-free (almost) features of electric vehicles are bound to win favor. The manufacturers must not compromise the inherent reliability of the electric vehicle with cost-cutting, but unreliable electronics.

- If electric vehicles get off to a good market start, a snowball effect known to economists as the "learning curve" will set in and costs will dramatically decline. Accompanying this will be the less tangible, but never absent, unexpected technological advances.

- In popular literature, the public is often influenced to expect a large energy contribution from solar cells appended to the vehicle. Unfortunately, even if solar cells could attain efficiencies of 50 percent (rather than 10–15 percent in mundane varieties) there is not sufficient surface area to gather in much energy. However, in some cases, a solar panel on the roof can provide some useful energy. It does not appear likely to become a common feature because of its fragility and vulnerability to dirt, and its reliance on the vagueness of weather.

- The public is also fed distorted information about the benefits of regenerative braking. It appears like perpetual motion in action because energy ordinarily expended to heat brake linings is, instead, directed back into the battery. It turns out that in most driving, relatively little energy is recovered. However, the possibility of increasing range by even a few percent will not be overlooked. Also, the mechanical brakes will last longer when preceded or assisted by electrical deceleration imposed by regenerative braking. Accordingly, this feature will be in evidence if for no other reason than its psychological appeal—obtaining a measure of "free" energy.

- Very likely, some type of hybrid format will also appear—one utilizing electrical propulsion assisted in some way by a small internal-combustion engine. In such a format, a worthwhile compromise might be realized where the most desirable features of electric and heat-engine propulsion are blended. It is conceivable that the gasoline-fueled engine would be used to charge, or to help charge, the battery, rather than be coupled to the drive shaft. It has also been proposed that a small gasoline engine could be used to actually power the wheels of the vehicle, but when a burst or increase of power was needed for acceleration, passing, or hill climbing, an electric motor would be energized from the on-board battery.

- There is likely to be *both* overnight-charging at home, and a number of strategically located quick-charge stations. Also, for safety reasons, charging energy can be supplied to the vehicle battery via inductive pickup.

- Advertising hype commonly omits mention that operation of amenities, such as vehicle heating or air conditioning, necessarily subtracts from otherwise available range.

- Although the knee-jerk reaction is to reject battery formats that have to operate at elevated temperatures, such as the sodium-sulfur type, such electrochemical systems posses an important feature denied to batteries that are said to operate at ambient temperatures. The very fact that elevated temperature batteries must be provided with superb thermal insulation to prevent heat leakage to the environment endows such a system with a remarkable feature; the same thermal insulation that inhibits heat leakage from the exotic battery also immunizes the battery from the effects of wide ambient temperature swings. Thus, such batteries will suffer minimal performance degradation in both, frigid and torrid climates. This feature came "along with the ride" in the thermal design primarily intended to inhibit battery heat from escaping.

- When one evaluates information on new battery systems, the most useful data is that expressed in terms of how the parameters compare to those of the lead-acid battery. Thus, if we read that, compared to lead-acid batteries, newly developed battery "X" has twice the specific energy and one and one half times the specific power, but only a quarter of the life span measured in charge/discharge cycles, we gain a practical idea of the new battery's merit, especially if some relative costs accompany such comparisons. However if our data is couched in terms of BTU per pound, Kwh/Kg, or Joules per cubic-inch, chances are good that we will wind up comparing apples and oranges. Inconsistent use of energy and energy-density units, as well as speed and range, loom as the major reason for much conflicting comparisons of batteries. Sometimes these erroneous evaluations are rather subtle, such as erratic and careless use of miles per hour and kilometers per hour. Unfortunately, these inconsistencies have a habit of passing the scrutiny of the author, the editor, and the reader!

- Admittedly, many statements invested with technical integrity at the time of writing, stand destined to best be taken "with a grain of salt" after passage of more time. This is due to the state of flux electric vehicle technology finds itself in. Experience teaches that a rapidly evolving applied science is full of unanticipated surprises. An abrupt breakthrough in a related device, such as electrolytic capacitors, or an unsought phenomenon in another energy investigation, such as the controversial "cold fusion", could dramatically impact electric vehicle technology. Indeed, once the electric vehicle gains a foothold in the market, its further progress will appear to feed on itself.

- Finally, "far-out" technology some of which we have only had tantalizing glimpses, might provide future mutations of the electric-vehicle art. These include high-temperature super conductivity, conducting polymers, 90%-efficient solar cells, replacement of electrochemical cells by rotating flywheel energy-storage devices, association of metal-hydride storage gas with electric motor propulsion, and the like. Note that we do not base our prognostications on UFO phenomena, but rather on projections of *known* principles. Two of these possibilities are illustrated in Figs. 8-8 and 8-9.

The amount of mechanical energy that can be stored in a rotating flywheel is surprising. Feasibility tests indicate specific energy levels of at least several times that of lead-acid storage batteries, with possibilities of even higher values. The simplified drawing of Fig. 8-8 depicts the basic setup underlying a practical implementation. The space constraints imposed by an electric car prevents the use of a

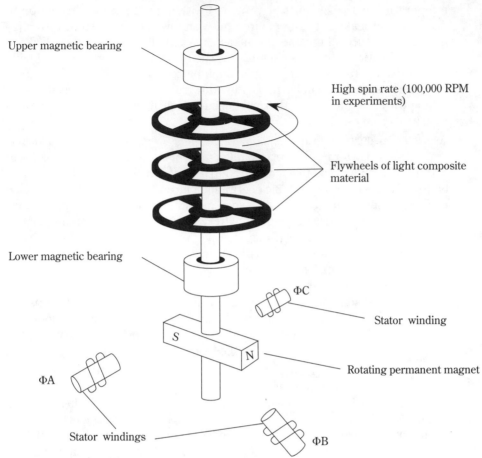

Upper magnetic bearing

High spin rate (100,000 RPM in experiments)

Flywheels of light composite material

Lower magnetic bearing

ΦC

Stator winding

Rotating permanent magnet

ΦA

S

N

Stator windings

ΦB

8-8 Possible mechanical "storage battery" of the future. Energy is stored in the high-speed spin of flywheels. This energy is proportional to the *square* of the speed. The rotating magnet and the externally mounted stator windings serve as both, a brushless dc motor for imparting the spin, and as a generator for supplying electrical energy to a load once the system has been "charged".

large-diameter flywheel. Also, traditional flywheels made of heavy metal tend to "explode" from centrifugal force if spun too fast. The solution to this problem is to use several relatively small wheels rotated at great speed. To be sure, kinetic energy is lost with the reduced diameter, and also because a light composite material is used. However, much of the lost energy can be regained if the speed of rotation is great enough. This is because the energy increases as the *square* of the speed. The composite material provides a safety feature, for its failure mode is more like a collapse than an explosion.

Magnetic bearings are used (details not shown), and the structure is housed in an evacuated chamber. Thus, friction and windage losses are exceedingly low; once "charged" to a high spin rate, the energy of rotation tends to maintain itself. It is like having a battery with a very low internal current leakage. A permanent magnet is also carried on the rotating shaft so that electricity can be generated in externally mounted

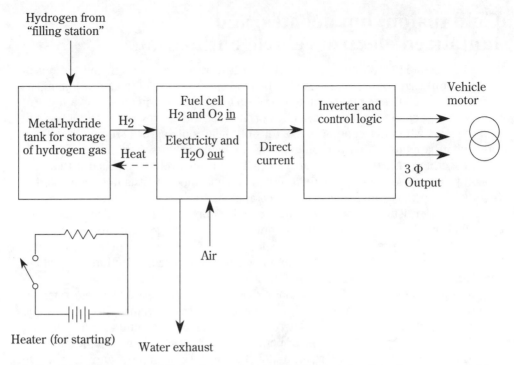

8-9 Possible energy format of the future electric vehicle. Certain alloys known as metal hydrides can absorb massive amounts of hydrogen gas in a way somewhat analogous to the way a sponge soaks up a liquid. This appears to be more practical and safer than storing the hydrogen under pressure, or as a liquid.

stator windings. With suitable electronics, this assemblage can also function as a motor; in this way, the system can be "charged" by bringing the flywheels up to speed.

Another "high-technology" format for an electric vehicle is shown in the simplified drawing of Fig. 8-9. Here, hydrogen gas is stored in an unusual container. It is then fed to a fuel cell where it mixes with incoming oxygen from an air intake, and thereby produces electricity, while exhausting hot water. Of course, the electrical energy can be utilized in various ways. In the drawing, a three-phase inverter is used in order to supply power to an induction motor, a synchronous motor, or a brushless dc motor. The heart of this scheme lies in the relatively safe means of storing the hydrogen gas. The hydrogen is not simply "contained" in the special alloy tank. Rather, it is *absorbed* in interstitial microscopic spaces in the metallic crystal structure. It is a phenomenon hard to imagine in terms of everyday experiences at the macroscopic level, but more hydrogen can be stored in this way that could be carried by a like-sized tank filled with liquefied hydrogen. (A workable, but unsafe storage technique). Such alloys are known as metal hydrides.

For the hydride to release its absorbed hydrogen atoms, it must be heated above the ambient temperature. Once such an electric vehicle was operative, it is likely that this could be accomplished with the hot-water exhaust from the fuel cell. Initially, however, heat would have to be obtained from an auxiliary source, such as lead-acid batteries carried along for the specific tank of starting the system.

Cold fusion, hot debates, and tantalized electric vehicle enthusiasts

Probably second in fervor only to the UFO controversies have been the allegations of "cold-fusion" that some people claim to be readily achievable with simple apparatus bearing resemblance to an experimental setup for demonstrating electroplating. Fusion just happens to be the process responsible for the energy released by the sun and the stars wherein temperatures of tens-of-millions of degrees are involved. Basically what happens is that hydrogen atoms are caused to fuse together to form helium, together with the release of energy. It is a much cleaner process than the fission process used in our atomic utility plants, inasmuch as it is not accompanied by much of the toxic and radioactive substances associated with atomic fission. Accordingly, scientists have been seeking methods of producing controllable fusion for a number of decades. Their general technique has been to generate extremely high temperatures and pressures in a small pellet of material for very brief time durations. Very expensive laser equipment, electromagnets, and instrumentation has yielded encouraging, but not practical results.

You can imagine the consternation of these pursuers of "establishment science" to awake one morning and read about fusion demonstrated via the use of a simple electrolytic cell. On the one hand, there seemed to be valid grounds for denouncing the claims as a hoax of the caliber of periodic "inventions" of perpetual motion. On the other hand, there surely was an unholy blend of "sour grapes," and "how in the world did I overlook this discovery?" Amidst gnashing of teeth, charges, and counter chargers, scientists, physicists, chemists, and home experimenters reported a confusing array of results. The excitement over the matter is easy enough to account for; after all, the potential energy in a cubic foot of sea water is about equivalent to that in 10 tons of coal. The mention of sea water is appropriate because it is made up not only of "ordinary" heavy water is formed of deuterium and oxygen atoms. Deuterium is an isotope of hydrogen, identical in its chemical properties, but containing an extra neutron in its atomic nucleus.

In the initial claim of cold fusion, current from a battery was passed through a cell comprised of palladium electrodes immersed in a heavy water electrolyte. Measured heat evolvement from the cell exceeded that which should have been generated from purely chemical reactions. Also, it was thought that helium and neutrons were given off from the cell—an accepted proof of nuclear phenomena. The general idea might be gleaned from Fig. 8-10.

Later, other experimenters confirmed the excess-heat observation, but not many detected helium or neutrons. This led to a popular consensus that an unusual *chemical* reaction might be involved, but there was also the assertion that most investigators were guilty of sloppy heat measurement techniques. This school of thought maintains that very thorough mixing of the liquid in a vessel is necessary before attempting to record temperature rise.

Whether or not otherwise-competent scientists possess sufficient sense to properly agitate a liquid bath in order to uniformly spread its temperature is a mute point. And whether any excess heat comes about because of atomic fusion or because of enhanced chemical activity, the practical relevance to electric vehicles demands further investigation. The question that must be answered is whether the process of en-

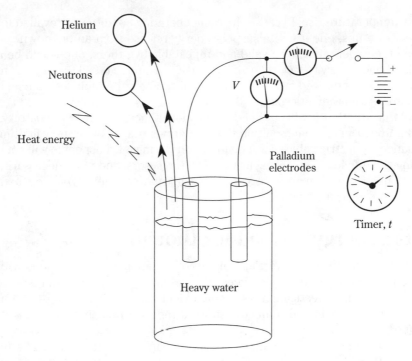

8-10 Basic setup for seeking evidence of cold fusion. Validity of concept requires three confirmations: 1. Heat energy evolved must exceed electrical input energy, $V \times I \times t$. 2. Helium atoms must be released from the heavy water electrolyte. 3. Neutrons must be released from the heavy water electrolyte. If only the first confirmation can be made, the phenomenon commands interest as some type of enhanced chemical reaction

ergy generation, if it indeed exists, can be harnessed to produce electricity on a greater energy-to-weight ratio than has been possible, practical, or economic from mundane and exotic batteries and fuel cells. The tantalizing aspect of the matter is that we are again dealing with an electrolytic cell, one that is all too suggestive of both voltaic and fuel cells.

Adding to the mystery of the claims is the use of palladium electrodes. This metal has been long used as a catalytic agent, a substance that speeds chemical reactions without itself undergoing chemical change. This is interesting because catalytic action, like gravitation and magnetism although well enough understood to manipulate its use in our technology, its true nature continues to be a fuzzy enigma. Fuel cells also are heavily dependent on the presence of a catalytic metal. It should not be an exercise in idle speculation to ponder why the two devices share this common feature.

Aside from the theoretical arguments of the physicists and chemists, a suggestive experiment was carried out at the Stanford University engineering department. Two identically arranged systems were monitored for the presence of excessive heat. In one, however, ordinary water was used; in the other, heavy water was used. It was found that heat evolvement from the cell with ordinary water was possible to ball park from computations. This was construed to be the expected "chemical" heat. The heavy-water cell yielded considerably more heat energy as manifested by a greater temperature rise. Because the same operating conditions prevailed for the side-by-side setups, one is led to suspect energy release from an unknown source. Assuming the experiment to be valid and repeatable by others, one would be more inclined to bend theory to fact, than vice versa. It is fitting to recall that a great mathematician once announced heavier-than-air flight to be impossible (of course, bumble bees are not mathematicians).

I feel that whether it will be called cold fusion, a fuel cell, an exotic battery, or a trillion-farad capacitor, some solution to the energy-storage barrier now limiting the performance of electric vehicles will be found. In the meantime, people will continue upgrading the efficiency and capability of electric motors, and must necessarily pay inordinate heed to such matters as vehicle weight, payload, aerodynamic streamlining, and tire-to-road traction.

"Horse energy"—food for thought

Our digression into the controversial "cold fusion" phenomenon, concluded with the suggestion that even if the alleged excessive-heat energy was not from fusion at all, it merited further investigation as a high-yield chemical reaction—one more exothermic than anticipated from conventional theory. I find that this leads to another interesting speculation.

Consider the strength, endurance, and work capacity of the horse. The fuel of this biological machine is simply grass or hay. Chemical oxidation of this fuel would seemingly provide only a small fraction of the horsepower-hours of work the noble steed performs in the service of man. Where, indeed, does this animal's energy come from? From the hay, to be sure, but via the digestive process under the influence of enzymes. And what are enzymes but biological catalysts performing similar tasks to the inorganic catalysts in more ordinary chemical reactions, and in fuel cells, automobile exhaust converters and apparently, the palladium electrodes of cold-fusion cells? Incidentally, the horse is not unique—the migratory flights of birds over continents and oceans wherein they might be buffeted by raging storms and impeding winds also exemplify this mystery. Here again, the derivation of the required energy surely does not come from bugs, worms, and seeds in the form of simple chemical breakdown. Nature harbors a secret here; can we find it?

Because this book is primarily devoted to motor and control technology, these speculations of mysterious energy sources will not be further expounded. The brief mentions that have been made are, of course, relevant to the future of the electric vehicle. It is encouraging that the sought energy-storage device need only surpass the performance parameters of the lead-acid battery by factors of three, four, or five in order to provide ranges of several-hundred miles at freeway speeds, and with practical payloads. Although, battery improvement has been somewhat disappointing, it nonetheless appears that too much magic is not needed; a favorable blend of technical, economic, and psychological factors is bound to bring about wider use of electric automobiles. It is certainly true that the electric motors and control techniques we have investigated are reasonably ready for the task.

Appendix
Useful information

Speed, slip, and frequency
Speed of any dc motor-general equation

$$S = \frac{V_a - (I_a R_a + E_b)}{k\,\Phi}$$

where,
  S is the speed of rotation in RPM,
  V_a is the voltage applied to the armature terminals,
  I_a is the armature current in amperes,
  R_a is the armature resistance in ohms,
  E_b is the voltage drop across the brushes,
  Φ is the flux lines per pole linking the armature conductors,
  k is the constant for any given machine.

$$k = \frac{ZP \times 10^{-8}}{60a}$$

where,
  P is the number of poles,
  Z is the number of active conductors on the surface of the armature,
  a is the number of parallel current paths in the armature winding.

Speed of a given dc series motor

$$S = K\left[\frac{V_a - I_a(R_a + R_e) - E_b}{I_b}\right]$$

where,
  S is the speed of rotation in RPM,
  V_a is the voltage applied across the motor,

I_a is the motor current in amperes,
R_a is the resistance of the series field in ohms,
R_e is the resistance of the armature in ohms,
E_b is the voltage drop across the brushes,
K is a constant for a particular machine.

$$K = \frac{I_a}{k\,\Phi}$$

Speed of any dc series motor

$$S = \frac{V_a - I_a(R_a + R_s) - E_b}{k\Phi}$$

where,
  R_s is the resistance of the series field in ohms.

Note: See general equation for description of other parameters.

Speed of any dc differential compound motor

$$S = \frac{V_a - I_a(R_a + R_s) - E_b}{k(\Phi_f - \Phi_s)}$$

where,
  R_s is the resistance of the series field in ohms,
  Φ_f is the shunt field flux,
  Φ_s is the series field flux.

Note: See general equation for description of other parameters.

Speed of any dc cumulative compound motor

$$S = \frac{V_a - I_a(R_a + R_s) - E_b}{k(\Phi_f - \Phi_s)}$$

where,
  R_s is the resistance of the series field in ohms,
  Φ_f is the shunt field flux,
  Φ_s is the series field flux.

Note: See general equation for description of other parameters.

Synchronous speed

$$S = \frac{f \times 120p}{P}$$

where,
  S is the speed of rotor in synchronous motor, or the rotating field in induction motor, in RPM,
  f is the frequency of the applied ac voltage in hertz,

P is the number of poles in the stator winding,
p is the number of phases.

Subsynchronous speed

$$S = \frac{f \times 120}{n}$$

where,
S is the speed of rotor of subsynchronous motor in RPM,
f is the frequency of the applied ac voltage in hertz,
n is the number of teeth on the rotor.

Note: The number of stator poles, or phases, does not govern the subsynchronous speed.

Rotor speed of induction motor

$$N_2 = N_1(1 - s)$$

where,
N_2 is the rotor speed,
N_1 is the synchronous speed (speed of the rotating field),
s is the slip expressed as a decimal.

Note: For a synchronous motor, $s = 0$ and $N_2 = N_1$.

Slip of rotor in an induction motor

$$s = \frac{N_1 - N_2}{N_1}$$

where,
s is the slip speed expressed as a decimal,
N_1 is the synchronous speed (speed of the rotating field),
N_2 is the actual speed of the rotor.

Note: *Percentage slip* $= s \times 100$.

Frequency of an alternator

$$f = \frac{PS}{2} = \frac{P \times revolutions\ per\ minute}{120}$$

where,
f is the frequency in hertz,
P is the number of poles,
S is the speed in revolutions per second.

Rotor frequency in an induction motor

$$f_2 = sf_1$$

where,
f_2 is the frequency of the current induced in the rotor,
s is the slip speed expressed as a decimal,

f_1 is the frequency of the stator current.

Note: For a synchronous motor, $s = 0$. Therefore, $f_2 = 0$.

Force and torque
Force exerted on an armature conductor

$$F = \frac{BI1 \times 10^{-7}}{1.13}$$

where,

F is the force exerted on the armature conductor in pounds,
B is the flux density of the field in lines per square inch,
I is the conductor current in amperes,
1 is the length of the conductor in inches.

Note: This equation assumes an orthogonal relationship between field and conductor.

Shaft torque of motors

$$T_{\text{out}} = T_a - T_r$$

where,

T_{out} is the torque available at the shaft of the motor,
T_a is the torque developed in the armature or rotor,
T_r is the torque dissipated by rotational losses.

Note: T_r comprises bearing friction, windage loss from spinning armature or rotor, windage loss from the fan, and brush friction.

Torque developed in armature of any dc motor—general equation

$$T_a = \frac{0.117(P)ZI_a \, \Phi \times 10^{-8}}{a}$$

where,

T_a is the torque in lb.-ft. developed in armature,
P is the number of poles,
a is the number of parallel current paths in the armature,
Z is the number of active conductors in the armature,
I_a is the total armature current in amperes,
Φ is the flux per pole linking the armature conductors.

Torque developed in armature of a given motor

$$T_a = k \, \Phi I_a$$

where,

T_a is the torque in lb.-ft developed in the armature,

$$k = \frac{0.117\,(P)Z \times 10^{-8}}{a}$$

Note: See general equation of armature current for description of other parameters.

Polyphase relationships
Relationships in two-phase motor operating from 3-wire line

$$E_d = \sqrt{2}(E_p)$$

where,
  E_d is the voltage across the stator winding,
  E_p is the voltage measured from the center tap of the stator winding to either side of the line.

Power delivered to three-phase motor

$$P = \sqrt{3}(E_L)(I_L)\mathrm{Cos\Theta}$$

where,
  P is the total power delivered to the three-phase motor
  E_L is the line voltage measured between any two phases,
  I_L is the current in a single line,
  Φ is the angle between voltage and current in any phase. $\mathrm{Cos\Theta}$ is the power factor.

Note: This equation applies to both balanced-Y and balanced-delta connections of the stator windings.

Power factor of three-phase motor

$$P_F = \frac{P}{\sqrt{3}(E_L)(I_L)}$$

where,
  P_F is the power factor,
  P is the power delivered to the motor,
  E_L is the line voltage between any two phases,
  I_L is the line current.

Note: Assumption is made that the motor presents a balanced load to the line.

Generators and alternators
Generator action

1. A coil of one turn has one volt induced when the rate of change of flux threading through is 10^8 lines per second.
2. The average EMF, E, developed in a single-turn coil is given by the equation:

$$E = \frac{\Phi \times 10^{-8}}{t}$$

where,

E is the average EMF developed in the coil,

Φ is the field flux,

t is the time it takes the coil to move from Φ field flux to zero field flux.

3. In a generator, the average value of the terminal voltage is given by the equation:

$$E_T = \frac{\Phi PZs \times 10^{-8}}{P}$$

where,

E_T is the average terminal voltage,

Φ is the number of flux lines extending from each pole,

P is the number of poles,

Z is the total number of armature conductors,

s is the speed of the generator in revolutions per second,

p is the number of parallel paths in the armature winding.

EMF induced in a conductor cutting a magnetic field

$$e = B1v \times 10^{-8}$$

where,

e is the induced EMF,

B is the flux density in lines per square centimeter,

1 is the length of the conductor in centimeters,

v is the relative velocity between the flux and the conductor in centimeters per second.

Note: An orthogonal relationship is assumed between B, 1, and v.

Instantaneous voltage in a coil rotating in a uniform magnetic field

$$e = E_m \, \text{Sin} \omega t$$

where,

e is the instantaneous voltage at time t,

E_m is the maximum instantaneous voltage,

ω is the angular velocity $2\pi f$ (f is the frequency in hertz),

t is the time in seconds.

Counter EMF in dc motors

$$E_c = V_a - (I_a R_a + E_b)$$

where,

E_c is the counter EMF in volts,

V_a is the voltage applied to armature terminals,

I_a is the armature current in amperes,

R_a is the armature resistance in ohms,

E_b is the voltage drop across the brushes.

Effective EMF per phase in an alternator

$$E_p = 2.22Z_s\,\Phi f \times 10^{-8}$$

where,

E_p is the effective sinusoidal voltage developed per phase,
Z_s is the number of conductors in series per phase,
Φ is the total flux cut per pole,
f is the frequency in hertz.

Note: This formula is valid for a *concentrated* winding. For a distributed winding, the conductor emfs do not add arithmetically.

Current and magnetic field relationships
Hand rule for current-carrying conductor

Grasp conductor with the fingers of your left hand, thumb extended parallel to the conductor. If your fingers indicate the direction of the flux encircling the conductor, your thumb points in the direction of the electron current flow. If your thumb points in the known direction of current flow, your fingers then indicate the direction of the magnetic flux encircling the conductor.

Hand rule for coils or solenoids

Grasp the coil with the fingers of your left hand, thumb extended parallel to the axis of the winding. If your fingers indicate the direction of electron flow, your thumb then points to the north pole.

Hand rules for motors and generators

For both machines, your thumb, forefinger, and middle finger should be positioned in a mutually orthogonal relationship. For example, your right thumb could point upward, your right forefinger could point West, and your right middle finger could point South. A 90° angle would then prevail between the planes encompassing any two of the indicated directions.

Motors: use your right hand—your middle finger points in the direction of electron flow. Your forefinger points in the direction of the field. Your thumb points in the direction of the resultant motor motion.

Generator: use your left hand—your thumb points in the direction of the moving conductor(s). Your forefinger points in the direction of the field. Your middle finger points in the direction of the resultant induced electron current.

Note: In both cases, the lines of the magnetic field are assumed to emerge from the north pole.

Miscellaneous
Efficiency of any motor

$$Eff. = \frac{P_{OUT}}{P_{IN}}\ \frac{P_{IN} - P_{LOSS}}{P_{IN}}$$

where,

  Eff. is a ratio that approaches unity as losses are diminished,
  P_{OUT} is the power available at the motor shaft,
  P_{IN} is the power consumed by the motor,
  P_{LOSS} is the internal power losses of the motor.

Note: P_{OUT}, P_{IN}, and P_{LOSS} must be expressed in the same units of power, such as watts. Eff. is converted to a percentage by multiplying by 100.

Current and torque relationship in motors

In order to develop maximum torque and minimum heating, the current waveform should have a high

$$\frac{I_{AVG}}{E_{EFF}}$$

ratio. In this expression, I_{AVG} is the average value of current in the armature or rotor, and I_{EFF} is the effective value of current in the armature or rotor.

Effective value of sine wave
Voltage induced in a transformer winding

$$E = 4.44fNB_{max}A \times 10^{-8}$$

where,

  E is the effective value of induced voltage (assumes operation from a sine-wave source),
  f is the frequency in hertz,
  N is the number of turns,
  B is the maximum flux density in lines per square inches,
  A is the cross sectional area of the core in square inches.

Note: Transformer theory is often applicable to the induction motor where the stator is the "primary" and the rotor is the "secondary."

Field current in a dc shunt motor

$$i = \frac{V}{R_t}\left(1 - e - \frac{t}{R_f L_f}\right)$$

where,

  i is the dc field current at time t after energization,
  V is the dc voltage applied to the field,
  R_f is the resistance of the field in ohms,
  L_f is the inductance of the field in henrys,
  t is the time after the field is energized in seconds,
  e is 2.718 (the base of natural logarithms).

Index

O

Oerstead, Hans Christian, 2
off/on control, triac, three-phase
    induction motor, 148, **150**,
    151

P

permanent magnet motors, 60,
    60, 62-64, **62**, **63**
  Alnico magnets, 62
  demagnetization effects, 63
  materials used, 62
  repulsion ac motor, **70**, 71
  speed controls
    feedback circuit, 113-116,
      114, **115**
    feedback system, 120, **122**,
      123
    synchronous, 123-126, **124**,
      125, **126**
    speed and torque, 63-64, **63**
    standstill point, 63-64, **63**
    starting torque, 64
    stepping, bipolar drive, 227-231,
      228, **229**, **230**
    stepping, unipolar drive, 222-
      227, **224**, **226**
    torque control, 119-120, **119**,
      120, **121**
phase-locked loop motor speed
    control, 129-132, **130**, **131**
photons and light theory, 2
poles of magnets, 6
polyphase relationships, 279
  three-phase motor, power deliv-
    ered, 279
  three-phase motor, power fac-
    tor, 279
  two-phase motor, three-wire
    line, 279
position control
  dc motors, speed-position con-
    trol, 136-142
power factors
  energy conservation, power fac-
    tor control, 179-185, **181**,
    182, **184-185**
  induction ac motors
    polyphase, 92
    repulsion-start, 86
    shaded-pole, 83
    split-phase, 75
    split-phase, capacitor-start, 78
    split-phase, permanent-capac-
      itor, 80
    split-phase, resistor-start, 75

  two-value capacitor, 81
  synchronous ac motors, 94
  three-phase motors, 279
power supplies
  dc, regulated, use as controller,
    203, **204**
Presley Co., 212
programmable divider speed con-
    troller, 205
propagation of forces in fields, 2
pulse-width modulated motor
    control, 126-129, **127**
pulse-width modulation
  ac motors, 172-173, **172**, **173**,
    174
  dc motors, 214, 216-217
  IC controller circuit, 212-213
  LM3524 regulating PWM IC,
    213-214, **213**, **214**, **215**, **216**,
    217

Q

quantum physics, 2

R

radiometer, 3-4, **3**
RCA, 176, 259
rectifier action of commutators,
    107
regenerative braking, 101-102,
    102
  electric automobiles, 268
  variable frequency inverters,
    102
regulated power supplies as mo-
    tor controller, 203, **204**
regulated speed control of univer-
    sal motor, 109-110, **109**
regulation, voltage, auto alterna-
    tor, 170-172, **171**
relativity theory, 2
reluctance, 28-29, **28**
  motor action, 28 29, **28**
repulsion ac motors, 67, 69-71
  brush/magnet configurations,
    69-70, **70**
  electromagnetic induction, 69
  permanent magnet arrange-
    ment, **70**, 71
  speed, 69-71, **71**
  torque, 69-71, **71**
  transformer action, 69
repulsion of magnets, 6, 14
resistance modulation speed con-
    trol, 201-203, **202**

reversal of rotation in dc motors,
    44
reverser, garage-door opener mo-
    tor reverser, 145-146, **145**
RFI generation
  commutators, shorted commu-
    tators, 36
  induction ac motors, repulsion-
    start, 86
rotator, antenna rotator, 196-199,
    197, **198**
rotor, 72
  frequency, induction ac motors,
    277
  induction ac motors
    frequency, 277
    rotor slip, 277
    rotor speed, 277
  slip, induction ac motors, 277
  speed, induction ac motors, 277
  synchronous ac motors, 93
    pole faces, 94
runaway region, dc motor speed
    control, 53, **53**

S

saturation, magnetic saturation,
    18
SCR speed control of universal
    motor, 107-108, **108**
SCR with gate turn-off capability,
    176-177, **177**
self-propagation of forces within
    fields, 2
series ac motors, 65-66, **67**
  characteristics and operation,
    65-66
  inductive compensating wind-
    ings, 65-66
  pulse-width modulated control,
    126-129, **127**
  reversal of rotation, 66
  space quadrature, 66
  transformer action, 66
  universal series ac motors, 66,
    68
series dc motors, 55-56, **56**
  efficiency of motor, 281-282
  generators use, 60-62, **60**, **61**
  horsepower, 58-59, **59**
  overload condition, 56
  shaft horsepower, 58-59, **59**
  speed control, 55-56, **56**
  speed rating, 58-59, **59**
  speed, 275-276
  speed-runaway feature, 56